BRICKWORK & BLOCKWORK

Dr Joseph Durkin

Member of the Chartered Institute of Building (MCIOB)
Construction and Work Based Learning
Hopwood Hall College
Rochdale
United Kingdom

This edition first published 2011
© 2011 by Joseph Durkin

Blackwell Publishing was acquired by John Wiley & Sons in February 2007.
Blackwell's publishing program has been merged with Wiley's global Scientific,
Technical and Medical business to form Wiley-Blackwell.

Registered office: John Wiley & Sons Ltd, The Atrium, Southern Gate, Chichester,
West Sussex, PO19 8SQ, UK

Editorial offices: 9600 Garsington Road, Oxford, OX4 2DQ, UK
The Atrium, Southern Gate, Chichester, West Sussex, PO19
8SQ, UK
2121 State Avenue, Ames, Iowa 50014-8300, USA

For details of our global editorial offices, for customer services and for information
about how to apply for permission to reuse the copyright material in this book
please see our website at www.wiley.com/wiley-blackwell.

Library of Congress Cataloging-in-Publication Data

Durkin, Joseph.
Brickwork and blockwork / Joseph Durkin.
p. cm.
Includes bibliographical references and index.
ISBN 978-1-4051-9977-3 (alk. paper)
1. Masonry. I. Title.
TH5311.D87 2011
693'.21–dc22

2010024204

A catalogue record for this book is available from the British Library.
Set in 9 on 13 pt Helvetica by Toppan Best-set Premedia Limited
Printed and bound in Singapore by Markono Print Media Pte Ltd

1 2011

CONTENTS

Introduction

This book has been written with the objective of assisting those students who are studying for the Level 1 Diploma in Bricklaying as well as those learners who are in employment and are working towards NVQ Level 1 Trowel Occupations.

The contents are based on and clearly mapped to the new syllabus for the Diploma in Bricklaying at Level 1 and should prove useful to both students and tutors alike.

Key features of the book

The features in the book have been written to increase and improve your knowledge and understanding of the Diploma and include the following.

Words and Meanings

New or difficult words and phrases will be highlighted in the text and a definition given.

Quick Quiz

There will usually be a quick quiz after each section. The quizzes are designed to test your knowledge of the work recently undertaken. They can be attempted at any time after reading the relevant section. If you do not do well first time, try again later.

Try This Out

'Try this out' activities will be of a more demanding nature than the quick quizzes and are designed to assist students who are preparing for internal and external exams at Level 1.

Brickwork and Blockwork by Joseph Durkin. © 2011 Joseph Durkin. Published 2011 by Blackwell Publishing Ltd

The contents of the questions and tasks are based on examinations of recent years and should prove useful for revision purposes.

Note: some of the tasks are to test practical ability and need to be carried out in a workshop or on site. It would be useful, therefore, if you liaise with your tutor about these.

Illustrations

Each page will contain clear and descriptive information about an activity or procedure.

Photographs

Many of the photographs that appear in the book have been specially taken and are designed to help you to follow a step-by-step procedure or identify a tool or material.

Useful Websites

At the end of each chapter or section there will be a number of useful websites for you to contact in order to further your knowledge of the topic area.

Sources of Further Information

At the end of each chapter or section there will be a number of sources of further information for you to contact in order to further your knowledge of the topic area.

Acknowledgements

I would like to thank the following for the help they gave me in preparing this book for publication: Joe Ellam, Paul Howells and Darren Whatmough for brickwork models and tool demonstrations; Danny Bowerbank and Bill Walker for IT support; Leigh Petrez and Mahmooder Khan for modelling PPE; John North of Bardsley Construction for generously allowing the site photography and lastly my wife, Caroline, for seeing it all through with me.

CHAPTER 1

Health and Safety

THIS CHAPTER RELATES TO UNIT CC1001K AND UNIT CC1001S.

Health and Safety Legislation

There are a lot of different pieces of legislation and regulations that affect you within the construction industry. Over the course of this chapter we will discuss just a few of the more relevant ones that you need to be aware of. Some will be dealt with in more detail later on in the work.

Health and Safety at Work Act 1974 (HASWA)

The HASWA covers the health and safety of almost everyone in the workplace. The main objectives of the HASWA are to:

- secure the health, safety and welfare of all persons at work
- protect the public from risk to their health and safety caused by work activities
- control the use, handling, storage and transportation of explosives and highly flammable substances
- control the release of noxious or offensive substances into the atmosphere
- ensure safety notices are displayed outside a building site.

The HASWA is enforced by inspectors, employed by the Health and Safety Executive (HSE). Their authority permits them to:

- enter premises to carry out investigations
- take statements
- check records
- give advice and information
- seize, dismantle, neutralise or destroy material, equipment or substances that are likely to cause immediate serious personal injury

Brickwork and Blockwork by Joseph Durkin. © 2011 Joseph Durkin. Published 2011 by Blackwell Publishing Ltd

- issue prohibition notices (ban all activity until the situation is corrected)
- issue improvement notices (put right within a specified period of time any minor hazard or infringement of legislation)
- prosecute all persons who fail to comply with their duty under the HASWA

Employers' and Employees' Duties under HASWA

Employers and employees have responsibilities under the HASWA. These are often referred to as 'duties' and are things that should or should not be done by law. If you do not carry out your duties, you may be breaking the law and you could be prosecuted.

Employer Responsibilities

All work activities are covered by health and safety law. The law which is most relevant to construction health and safety is the Health and Safety at Work Act 1974. This Act applies to all work activities.

It requires employers to ensure as far as reasonably practicable the health and safety of their employees, other people at work and members of the public who may be affected by their work.

Employers should have a health and safety policy. If they employ five or more people, the policy should be in writing.

Employers are required to plan, control, organise, monitor and review their work. To do this they should:

- assess the risks associated with work to identify the control measures necessary to reduce these risks
- have access to competent health and safety advice
- provide health and safety information and training to employees
- have arrangements to deal with serious and imminent danger
- cooperate in health and safety matters with others who share the workplace.

Employers must identify the hazards involved with their work, assess the likelihood of any harm arising and decide on adequate precautions. This process is called risk assessment and is central to all planning for health and safety.

Site personnel must be trained in safe working practices. Employees cannot be relied upon to pick up safety training on the job from their colleagues – they might simply be learning someone else's bad habits.

Employers must provide a safe place to work with safe entrances and exits and provide all site personnel with personal protective equipment (PPE) if required.

Summary

In summary, the employer must ensure that health and safety is taken into account and managed throughout all stages of a project, from conception, design and planning through to site work and subsequent maintenance and repair of the structure.

Employee Responsibilities

Employees also have health and safety duties. They should:

- follow instructions given to them by their supervisors
- cooperate with their employer on health and safety matters
- follow the health and safety rules which apply to their particular job and to the site in general
- use the health and safety equipment provided
- report defects in equipment to their supervisor
- take care of their own health and safety as well as that of their workmates and others who might be affected by their work.

Employees should be trained to know what to do, and the work should be supervised and monitored to make sure that information provided as training is relevant to the work situation and is applied effectively.

Control of Substances Hazardous to Health Regulations 2002 (COSHH)

The COSHH cover dangerous solids, liquids or gases and gives guidelines on how they should be used and stored (Figure 1.1). They give details of actions the employer and the employee must take to protect the health of the individual and others.

Using chemicals or other hazardous substances at work can put people's health at risk. So the law requires employers to control exposure to hazardous substances to prevent ill health. Employers have to protect both site personnel and others who may be exposed, by complying with COSHH.

Provision and Use of Work Equipment Regulations 1998 (PUWER)

The PUWER help to provide guidance to protect people's health and safety from equipment that they use at work.

The PUWER cover all working equipment such as tools and machinery. This equipment can include ladders, lifting equipment, earth-moving machinery, powered hand tools (Figure 1.2), cutting and drilling machines and so on.

Words and Meanings

Health and safety has a lot of terms and meanings that need to be read, understood and remembered.
Regulations – Rules that have been put in place to ensure work is carried out both correctly and safely.
Legislation – A law that has been made in Parliament and is often called an Act.

Figure 1.1 Storage of hazardous substances.

Figure 1.2 Power tools.

Under the PUWER, employers must make sure that any tools and equipment they provide are:

- suitable for the work to be carried out
- maintained, serviced and repaired on a regular basis
- inspected on a regular basis to ensure that the piece of equipment and its parts are in good working condition.

Employers also have to make sure that any risk of harm from using the equipment has been identified and that all precautions and safety measures have been taken. Employers must also ensure that anyone who uses tools and equipment has been properly trained and instructed in how to do so.

Manual Handling Operations Regulations 1992

Manual handling is transporting or supporting loads by hand or using bodily force. Many people hurt their back, arms, hands or feet lifting everyday loads, not just when the load is too heavy. More than a third of all over-three-day injuries reported each year to the Health and Safety Executive and to local authorities are the results of manual handling. These can result in those injured taking an average of 11 working days off each year.

Control of Noise at Work Regulations 2005

Excessive noise levels from site machinery can cause hearing impairment if exposed to it continually. Therefore ear defenders must be worn to protect the eardrums (Figure 1.3). Young people can be damaged as easily as the old, and premature deafness is even worse. Sufferers often first start to notice hearing loss when they cannot keep up with conversations in a group or when the rest of their family complain they have the television on too loud. Deafness can make people feel isolated from their family, friends and colleagues. (See the section on noise safety later on in this chapter.)

Electricity at Work Regulations 1989

The Electricity at Work Regulations cover any work that involves the use of electricity or electrical equipment. Employers have a duty to make sure that all electrical systems you may come into contact with are safe and regularly maintained.

Personal Protective Equipment at Work Regulations 1992 (PPER)

The use of personal protective equipment (PPE) is not the solution for preventing accidents; it is used as a last resort. It is of primary

Figure 1.3 Worker using noisy equipment.

Words and Meanings

Accident – An unplanned or unwanted event or occurrence that may result in injury to a person and/or damage to property.

importance to prevent accidents by identifying all possible hazards and then taking the necessary steps to eliminate them, *before* they cause an accident.

The PPER detail the different types of PPE that are available and state when they should be worn (Figure 1.4). For example, employers must provide employees with hard hats and make sure they are worn and worn correctly. (The different types of PPE available are covered in more detail later on in this chapter.)

Health and Safety (Safety Signs and Signals) Regulations 1996

These regulations bring into force the European Community Safety Signs Directive on the provision and use of safety signs at work. The purpose of the directive is to encourage the standardisation of safety signs throughout the member states of the European Union so that safety signs, wherever they are seen, have the same meaning. See the section on signs and notices later in this chapter for a closer look at the signs used on a construction site.

Reporting of Injuries, Diseases and Dangerous Occurrences Regulations 1995 (RIDDOR)

Figure 1.4 Some examples of PPE: high visual jacket, eye protection and hard hat.

Certain events must be reported to the HSE, for example major injuries and deaths that occur on a construction site or accidents in which employees are unable to work for more than three days as well as diseases and dangerous occurrences.

The Reporting of Injuries, Diseases and Dangerous Occurrences Regulations 1995 (RIDDOR) require that certain accidents that happen on site, listed in the following section, have to be reported.

Reporting Responsibilities and Accidents

Learners and employees in construction should be aware of the reporting system and should be able to report accidents verbally and understand the documentation used to record accidents. It is important that you are clear, accurate and factual about any accident you may have to report.

Remember Emergencies require immediate action.

Some examples of emergencies are fire, security alerts, uncontained spillage or leakage of chemicals or other hazardous substances, scaffold failure, bodily damage and health problems.

In the European Union, the construction industry is the industry most at risk from accidents, with more than 1300 people being killed in construction accidents every year.

Reporting Hazards and Injuries

Reporting accidents and ill health at work is a legal requirement. The information enables the HSE and local authorities, referred to as the enforcing authority, to identify where and how risks arise and to investigate serious accidents.

Record Keeping

You must keep a record of any reportable injury, disease or dangerous occurrence for three years after the date on which it happened. You can keep the record in any form you wish.

The Accident Book

All accidents, even minor ones, have to be entered into the Accident Book at work. The following must be recorded:

- the name of the injured person, their home address and occupation;
- the signature of the person making the entry, their home address, occupation;
- when and where the accident happened;
- a brief description of the accident, its cause and what the injury was;
- the method of treating the accident;
- other people involved;
- whether the accident is reportable to the HSE.

The HSE will also need to be informed if a major injury or dangerous occurrence takes place, or if the employee is off work for more than three days.

Location of Accident Book

There is no set place to keep an Accident Book. However, it needs to be easy to get at and often it is kept near the first-aid point.

Employers must tell employees where the Accident Book is kept. Sometimes employers just tell employees when they first start where it is kept. Sometimes notices are displayed. The company's safety policy may also say where it is.

Accident Trends within the United Kingdom

Each year, many building site workers are killed or injured as a result of their work; others suffer ill health, such as dermatitis, occupational deafness or asbestosis. The building industry's accident trends have shown a steady long-term improvement. However, the rates of death, serious injury and ill health are still unacceptably high.

Words and Meanings

Hazard – Something that can cause harm, illness or damage to health or property.

Words and Meanings

Accident – This is an unplanned or unwanted event or occurrence that may result in injury to a person and/or damage to property.

Note

The HSE introduced a new accident record book in May 2003. The new publication ensures companies comply with legal requirements to record accidents at work. It was revised in 2007 to take account of the requirements of the 2003 Data Protection Act.

Try this Out

Accidents have to be reported in writing using a form, an example of which is illustrated in Figure 1.5 (see elsewhere in this chapter for more information about accident reporting). Take a copy of the form and complete it, imagining that you have just witnessed an accident to a colleague.

Get your tutor to check that the form has been accurately completed.

---

1. About the person who had the accident

Name:_____

Address:_____

_____

Postcode: _____

Occupation:_____

---

2. About you, the person filling in this record

If you did not have the accident, write your address and occupation

Name:_____

Address:_____

_____

Postcode: _____

Occupation:_____

---

3. About the accident; (continue on the back of this form if you need to)

Say when it happened: Date: ___/___/_____   Time: ___:___

Say where it happened. State which room or place: _____

_____

Say how the accident happened. Give the cause if you can: _____

_____

_____

If the person who had the accident suffered an injury, say what it was: _____

_____

Please sign and date the record:

Signature:_____   Date: ___/___/_____

---

4. For the employer only

Complete this box if the accident is reportable under the Reporting of Injuries, Diseases and Dangerous Occurrences Regulations 1995 (RIDDOR)

How was it reported? _____

Date Reported: ___/___/_____   Signature: _____

Figure 1.5 Accident report form.

The Government and the Health and Safety Commission (HSC) set targets for improving the United Kingdom's health and safety performance. These targets include aims to:

- reduce the incidence rate of fatalities and major injuries by 66% by 2009/10;
- reduce the incidence rate of cases of work-related ill health by 50% by 2009/10;
- reduce the number of working days lost per 100 000 workers from work-related injury and ill health by 50% by 2009/10.

These targets will only be achieved if everyone involved in the construction industry plays their part.

Preventing Accidents

You can help prevent accidents by following the procedures, rules, training and instruction provided, and by cooperating with your employer.

Reporting to the Enforcing Authority

Reports have to be sent to the enforcing authority if there is an accident connected with work and:

- an employee, working on site is killed or suffers a major injury, or
- a member of the public is killed or taken to hospital.

You must notify the HSE without delay, usually by telephone. The HSE will ask for brief details about the injured person and the accident.

Within ten days, you must follow this up with a completed accident report form F2508 (Figure 1.6).

Over-Three-Day Injury

If there is an accident connected with work and an employee working on site suffers an over-three-day injury, you must send a completed accident report form (F2508) to the HSE within ten days.

An over-three-day injury is one which is not major but results in the injured person being away from work or unable to do the full range of their normal duties for more than three days.

Disease

If a doctor notifies you that an employee suffers from a reportable work-related disease, you must send a completed disease report form (F2508A) to the enforcing authority.

Words and Meanings

Major injury – Usually requires hospital treatment and a long time off work.
Minor injury – Usually does not require hospital treatment and does not involve more than three days off work.

Health and Safety at Work etc Act 1974 **?**
The Reporting of Injuries, Diseases and Dangerous Occurrences Regulations 1995

Click here for report guidance

HSE
Health & Safety
Executive

Report of an injury or dangerous occurrence

Filling in this form
This form must be filled in by an employer or other responsible person.

Part A

About you
1 What is your full name?

2 What is your job title?

3 What is your telephone number?

About your organisation
4 What is the name of your organisation?

5 What is its address and postcode?

6 What type of work does the organisation do?

Part B

About the incident
1 On what date did the incident happen?

2 At what time did the incident happen?
(Please use the 24-hour clock eg 0600)

3 Did the incident happen at the above address?
Yes ☐ Go to question 4
No ☐ Where did the incident happen?
☐ elsewhere in your organisation – give the name, address and postcode
☐ at someone else's premises – give the name, address and postcode
☐ in a public place – give details of where it happened

If you do not know the postcode, what is the name of the local authority?

4 In which department, or where on the premises, did the incident happen?

F2508 (05.00)

Part C

About the injured person
If you are reporting a dangerous occurrence, go to Part F. If more than one person was injured in the same incident, please attach the details asked for in Part C and Part D for each injured person.

1 What is their full name?

2 What is their home address and postcode?

3 What is their home phone number?

4 How old are they?

5 Are they
☐ male?
☐ female?

6 What is their job title?

7 Was the injured person (tick only one box)
☐ one of your employees?
☐ on a training scheme? Give details:

☐ on work experience?
☐ employed by someone else? Give details of the employer:

☐ self-employed and at work?
☐ a member of the public?

Part D

About the injury
1 What was the injury? (eg fracture, laceration)

2 What part of the body was injured?

Next Page

Figure 1.6   Report form F2508.

3 Was the injury (tick the one box that applies)

☐ a fatality?

☐ a major injury or condition? (see accompanying notes)

☐ an injury to an employee or self-employed person which prevented them doing their normal work for more than 3 days?

☐ an injury to a member of the public which meant they had to be taken from the scene of the accident to a hospital for treatment?

4 Did the injured person (tick all the boxes that apply)

☐ become unconscious?

☐ need resuscitation?

☐ remain in hospital for more than 24 hours?

☐ none of the above.

Part E

About the kind of accident

Please tick the one box that best describes what happened, then go to Part G.

☐ Contact with moving machinery or material being machined

☐ Hit by a moving, flying or falling object

☐ Hit by a moving vehicle

☐ Hit something fixed or stationary

☐ Injured while handling, lifting or carrying

☐ Slipped, tripped or fell on the same level

☐ Fell from a height

How high was the fall?

☐ _____ metres

☐ Trapped by something collapsing

☐ Drowned or asphyxiated

☐ Exposed to, or in contact with, a harmful substance

☐ Exposed to fire

☐ Exposed to an explosion

☐ Contact with electricity or an electrical discharge

☐ Injured by an animal

☐ Physically assaulted by a person

☐ Another kind of accident (describe it in Part G)

Part F

Dangerous occurrences

Enter the number of the dangerous occurrence you are reporting. (The numbers are given in the Regulations and in the notes which accompany this form)

For official use		
Client number	Location number	Event number
		☐ INV REP ☐ Y ☐ N

Part G

Describing what happened

Give as much detail as you can. For instance

- the name of any substance involved
- the name and type of any machine involved
- the events that led to the incident
- the part played by any people.

If it was a personal injury, give details of what the person was doing. Describe any action that has since been taken to prevent a similar incident. Use a separate piece of paper if you need to.

☐☐☐☐

Part H

Your signature

Signature

Date

Where to send the form

Incident Contact Centre, Caerphilly Business Centre, Caerphilly Business Park, Caerphilly, CF83 3GG. or email to riddor@connaught.plc.uk or fax to 0845 300 99 24

If returning by post/fax, please ensure this form is signed, alternatively, if returning by E-Mail, please type your name in the signature box

Continue

Figure 1.6   Continued

Dangerous Occurrence

If something happens that does not result in a reportable injury, but which clearly could have done, it may be a dangerous occurrence, which must be reported immediately by telephone to the enforcing authority. Within ten days, you must follow this up with a completed accident report form (F2508).

Accident Reporting

If you are involved in an accident or near miss at work:

- you may have to give details to your site manager, or
- you may be asked to fill in an accident report form as a witness or as an injured person.

These forms can be complicated. They must be read very carefully.

Accident and Emergency Records

Your employer has to inform you and all employees of the first-aid arrangements. Putting up notices telling employees who and where the first-aiders or appointed persons are and where the first-aid box is will usually be sufficient. Your employer also needs to make special arrangements to give first-aid information to employees with reading or language difficulties.

Working Safely

Serious accidents can easily happen in the construction industry, so keeping yourself and others safe at work is vital. For each job you do on site and in the workshop, you, your tutor and your employer must follow a process of risk assessment.

Remember You are legally required to assess the risks in the workplace.

The important things you need to decide are whether a hazard is significant and whether you have it covered by satisfactory precautions so that the risk is small. You need to check this when you assess the risks. See Figure 1.7, which shows some potential hazards that may find on a building site.

In most firms in the construction industry the hazards are many and complex. Checking them is common sense, but necessary.

Communicating Health and Safety

The HASWA is a very important piece of safety legislation. It is displayed on a poster (Figure 1.8) in every place of work and should be read and understood by all employers and employees. However, the quantity and difficulty of the text can make the task daunting.

Figure 1.7 Potential hazards.

Every year building workers are injured or killed at work. You need to be aware of what your employer should be doing to protect everyone's safety. You should also be clear about what safety areas you and your colleagues are responsible for (see the section on the HASWA above and in particular the responsibilities of employers and employees).

Figure 1.8 Example of a poster.

The Health and Safety Executive (HSE)

The HSE enforces the law in the workplace. It has the powers to inspect premises and construction sites to ensure employers and employees are not breaking the law.

One of the jobs of health and safety inspectors is to see how well site hazards are being dealt with, especially the more serious ones that could lead to injuries or ill health. They may also wish to investigate an accident or complaint.

HSE inspectors have wide powers, which include:

The right to issue an improvement notice, which gives a company a certain amount of time to deal with a health and safety issue.

The right to issue a prohibition notice (particularly where a risk of serious personal injury exists), which stops a process or the use of dangerous equipment.

Health and Safety on Site

Every year in the United Kingdom the HSE faces the challenge of reducing:

- about 350 fatalities to workers and members of the public due to reportable accidents at work
- some 12 000 early deaths due to past exposure to hazardous agents (e.g. asbestos)
- over 36 million working days lost due to work-related accidents and ill health.

Enforcement of Health and Safety

Health and safety laws which apply to construction companies are usually enforced by an inspector from the HSE. However, some smaller jobs inside offices, shops and similar premises are the responsibility of inspectors from the local authority.

Induction Programmes

On starting work or a college course you will be give an induction briefing session explaining your roles and responsibilities in the organisation. These will be backed up on a regular basis by what are known as tool box talks. These talks are done informally and are usually related to health and safety. It is important that you listen and take notice of them.

Construction Skills Certification Scheme (CSCS)

'Quality up, accidents down and cowboys out' were the main objectives for the formation of the Construction Skills Certification Scheme (CSCS) when it was launched in 1995.

Today, its main purpose is to help people who work in construction prove that they are competent to do their job and that they have health and safety awareness.

The levels of card (Figure 1.9) available are:

- CSCS cards list the holder's qualifications and are valid for one, three or five years.

Red
Working towards
N/SVQ level 2 or 3
Trainee

Gold
N/SVQ level 3*
Advanced Craft/
Supervisory

Yellow
Professionally
Qualified Person

Green
N/SVQ level 1 OR
Sign-off from employer
Site Operative (labourer)

Platinum
N/SVQ level 4
Management

Blue
Working towards
N/SVQ level 2 or 3
Experienced Worker

Black
N/SVQ level 5
Senior Management

White
N/SVQ
level n/a
Construction
Related
Occupation

Blue
N/SVQ level 2*
Craft

Yellow
N/SVQ level n/a
Visitor (no
construction skills)

Figure 1.9   Construction Skills Certification Scheme cards.

- The cards are based on the N/SVQ system. There are different coloured cards for each of the levels.

The back of each CSCS card states in which occupation(s) the cardholder is qualified.

To apply for a CSCS card, you must pass the Construction Skills' health and safety test.

Health and Safety Requirements

All applicants must take and pass an independent test of retained health and safety knowledge. The test is designed to examine the knowledge of the individual across a wide range of health and safety topics.

Construction Health and Safety Test

The Construction Skills construction health and safety test is available at four levels: operative, supervisory, management and professionally qualified person.

Authorised Persons

An authorised person is someone chosen to:

- take charge when someone is injured or falls ill, including calling an ambulance if required;
- look after the first-aid equipment (e.g. restocking the first-aid box).

Authorised persons should not attempt to give first aid for which they have not been trained, though short emergency first-aid courses are available. Remember that an authorised person should be available at all times people are at work on site – this may mean having more than one authorised person.

Authorised persons include:

- first-aiders
- supervisors
- safety officers
- managers
- members of the emergency services or HSE.

First-Aiders

An employer has to make adequate arrangements to treat employees and others who are injured or become ill at work. Employers have to appoint one first-aider, depending on the number of employees and the risk involved in the employer's work.

For low-risk employers and/or where there are few employees, it is adequate to authorise someone to take charge of a situation when someone has a serious illness or accident. Employers with a first-aider or first-aiders also tend to have these appointed persons to cover first-aid situations on sites that do not provide permanent first-aid staff.

First-Aid Boxes

People at work can suffer injuries or fall ill. It does not matter whether the injury or the illness is caused by the work they do. What is important is that they receive immediate attention and, in serious cases, an ambulance is called.

The Health and Safety (First Aid) Regulations require employers to provide adequate and appropriate equipment, facilities and personnel. This is so that employees can get first aid if they are injured or become ill at work.

What is adequate and appropriate will depend on the circumstances in your workplace.

Words and Meanings

First Aid – This is the method of treating minor injuries where other treatment is not needed. This minimises any chance of further injury or illness until a doctor, nurse or paramedic arrives.

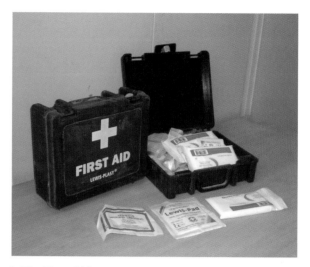

Figure 1.10   First-aid box.

The minimum first-aid provision on any building site is:

- a suitably stocked first-aid box (Figure 1.10);
- an appointed person to take charge of first-aid arrangements.

It is important to remember that accidents can happen at any time. First-aid provision needs to be available at all times when people are at work.

Employers must tell employees about first-aid arrangements at work. They usually do this by putting up notices telling staff who and where the first-aiders or appointed persons are, and where the first-aid box is.

Your tutor during the first week of your course will explain what first aid is and where the nearest first-aid box is located. Locate the first-aid box and make a list of its contents then ask your tutor for comments.

First-Aid Kits

An employer has to provide first-aid materials to deal with minor injuries or conditions such as cuts, fractures, burns and so on. A basic first-aid kit (Figure 1.11) should contain the following:

- a leaflet giving general advice on first aid (e.g. HSE leaflet *Basic advice on first aid at work*);
- 20 individually wrapped sterile adhesive dressings (assorted sizes);
- 2 sterile eye pads;
- 4 individually wrapped triangular bandages (preferably sterile);

Try this Out

With the information you already have and using the library, Internet and textbooks:

- Analyse national statistics regarding key accidents trends within the UK building industry.
- Compile a report of your finding and show it to your tutor for assessment.

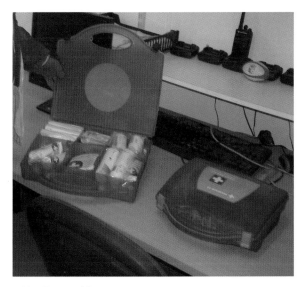

Figure 1.11 First-aid kit.

Figure 1.12 Tidy work area for the building of a large brick retaining wall, to hold back the embankment.

- 6 safety pins;
- 6 medium-sized (approximately 12 cm × 12 cm) wrapped sterile un-medicated wound dressings;
- 2 large (approximately 18 cm × 18 cm) sterile wrapped un-medicated wound dressings;
- 1 pair of disposable gloves.

You should not keep tablets or medicines in the first-aid box.

Hazards on Construction Sites

When you see a good bricklayer at work, the chances are the area in which he is working will be clean and tidy; quality work and a tidy workspace go hand in hand (Figure 1.12). This is what is meant by good housekeeping.

The Working Area

- Keep your work area tidy by stacking bricks correctly.
- Do not overload the mortar boards.
- Keep the tools you are not using in your tool bag.
- Leave time at the end of the day to clean the area and leave it ready for the next day.
- Sweep up debris.
- On site a brush can be used to sweep fine debris into heaps.

- In confined spaces or in the workshop, lay the dust by sprinkling water on the area to be swept.
- If dust cannot be avoided, always wear a dust mask.

Waste Bins

All brick and block waste should be deposited in waste bins for reuse or recycling.

A clean work area helps safety, improves working conditions and allows operatives to work efficiently.

Remember Materials are expensive – do not throw away good bricks and blocks.

Risk Assessments

Words and Meanings

Risk – Likelihood or chance that harm, illness or damage will occur and the degree of harm (how many people might be affected and how badly).

Risk Assessment – Mainly carried out by an employer to identify risks to his/her employees (and others) and to decide what is necessary to control these risks to the standards required under the law.

Workers on site should all personally assess the risk of working in particular settings or on particular jobs. At Diploma level 1 you will be recording risk assessment information; so you should be able to read and understand it.

This section covers the principles of risk assessment so that you can assess risk assessment information and take responsibility for your own health and safety.

Employers carry out formal risk assessments and write down safe working methods or method statements. You should also carry out a risk assessment for yourself every time you start a job.

There are three main steps to risk assessment:

- Step 1: Look for the hazards. Look for things that can cause harm or areas where accidents are more likely to happen.
- Step 2: Think about who is at risk from this hazard. Is it everyone or only workers doing one particular job? Are members of the public at risk?
- Step 3: What can be done to reduce the risk? Is there a safe method of working that will help to prevent accidents? Who is responsible for carrying out these safety precautions?

Method Statements

You need to be familiar with method statements, their purpose and their use on site. You need to make the link between method statements and risk assessments and the need to understand and comply with both.

Method statements may be written in a range of formats and will use some technical language. This may be a barrier to reading and understanding. Some method statements are quite lengthy and apply to a

Corfield Scaffolding LTD

Method Statement

Contractor: Buildfull

Site: Industrial Estate, Dursley, Gloucestershire

Date: 4th June 2005

1. **Package:** Scaffolding
2. **Description of Works:** Safety/crash deck
3. **Task:** Erect safety/crash deck
4. **Location:** Unit Tower 220
5. **Duration:** 1 day
6. **Labour Requirement:** 2 scaffolders
7. **Supervision:** An experienced foreman will supervise this task.
8. **Plant:** No mechanical plant will be used during initial erection.
9. **Materials:** All materials will be fully serviced and prepared for use. All scaffolding and fittings required for this task will be delivered from our yard to site as required.
10. **Management of the Work Area:** This scaffold will be erected within the site boundary. Other trades will not be working in the area.
    a) The perimeter of the work area will be cordoned off.
    b) Warning signs (scaffolding complete) will be posted at either end and remain in place until the job has been inspected at the end.
    c) No one will be allowed into the area unless they have a written permit from the foreman.
11. **Construction sequence:**
    a) Ensure contractor has prepared the base, which must be firm, level and clear of obstructions.
    b) All levels to be agreed with the site manager and checked.
    c) Erect the first board layer – to have two layers with plastic sheeting in between.
    d) Erect the second layer – board out using a minimum of three board runs for safety.
    e) Lay plastic sheeting before doing the top layer.
12. **Manual Handling:** All scaffolders are experienced and aware of safe handling of scaffolding materials.
13. **COSHH:** The freeing agent used on fittings is non-harmful even after prolonged exposure.
14. **PPE:** All scaffolders will wear and use hard hats, safety boots, hi-vis vests and body harness.

Figure 1.13 *Method statement, courtesy of the Department for Education.*

range of different teams. By scanning the headings of a document, you should be able to identify which parts of it are relevant to you.

Method statements are work instructions. They give you information about the job and procedures to follow, including safety measures. Make sure you read and understand all parts of the method statement before you start the job. See Figure 1.13.

Method statements vary from place to place. Make sure you are familiar with the ones you use.

You have a responsibility to look after your own safety and that of others on site. The method statement you have for your job tells you what will or should be done to make that job as safe as possible. This is why you need to read it in detail to make sure you understand it.

Near Misses

If something happens which does not result in a reportable injury, but which clearly could have done, it may be a near miss or dangerous occurrence, which must be reported immediately by telephone to the enforcing authority. Within ten days you must follow this up with a completed accident report form (F2508). Near misses and dangerous occurrences include:

- plant or equipment coming into contact with overhead power lines;
- electrical short circuit or overload causing fire or explosion;
- collapse or partial collapse of a scaffold over 5 m high, or erected near water where there could be a risk of drowning after a fall;
- malfunction of breathing apparatus while in use or during testing immediately before use;
- accidental release of any substance which may damage health.

Reporting near misses, accidents and ill health at work is a legal requirement. The information enables the HSE and local authorities, referred to as the enforcing authority, to identify where and how risks arise and to investigate serious accidents.

Hazards in the Workplace

On the majority of construction sites in the United Kingdom, there is great emphasis on protecting construction personnel and the public from hazards associated with the work area and certain types of work practice.

Falling Materials/Tools

Falling materials or tools on building sites can cause injury to the head and other parts of the body. It is not only essential: the law dictates protective headgear must be worn at all times by all site personnel and visitors in order to minimise the hazard.

Noise

Excessive noise levels from site machinery can cause damage to hearing if personnel are exposed to it continually. Therefore ear defenders must be worn to protect the eardrums (see Figure 1.3 above). (See the section on noise safety later on in this chapter.)

Manual Handling

Manual handling is transporting or supporting loads by hand or using bodily force. Many people hurt their backs, arms, hands or feet lifting everyday loads, not just when the load is too heavy. Most cases of injury can be avoided by providing suitable lifting equipment, which is regularly maintained, together with relevant training on both manual handling techniques and the safe use of equipment.

Handling and Storing of Materials and Equipment

When lifting loads without the use of lifting devices, care must be taken to avoid injury to the spinal column and the stomach and back muscles. By using the kinetic method of manual handling, injuries to the back can be avoided.

In your first week at college, you will be shown how to store tools, materials and equipment in the workshop. These may include, hand tools, hand-held power tools, wheelbarrows, ladders, trestles, scaffolding boards, brick, blocks and so on.

Electricity

Electricity can kill. Most deaths are caused by contact with overhead or underground power cables. Even non-fatal shocks can cause severe and permanent injury. Shocks from equipment may lead to falls from ladders, scaffolds or other work platforms. Those using electricity may not be the only ones at risk. Poor electrical installation and faulty electrical appliances can lead to fires, which can result in death or injury.

Fire Protection

Fire protection methods should be included in the organisation of any institutional or building site safety programme. The most common type of fire protection equipment is the fire extinguisher.

Fire extinguishers should be suitably placed, distinctly marked and easily accessible. Fire hoses, nozzles, connections, taps and pumps should be checked and maintained on a regular basis.

Enclosed workshops should have ample exits and all adjoining shops should be separated by fire walls or fire doors. Fire doors and walls provide a means of containing and preventing the spread of fire.

Work Equipment

Work equipment covers an enormous range, including process machinery, machine tools, lifting equipment, hand tools and ladders. Important points include:

- selecting the right equipment for the job;
- making sure equipment is safe to use and keeping it safe through regular maintenance;
- inspection and, if appropriate, thorough examination;

Note

Do not try to lift anything that weights over 25 kg on your own.

Try this Out

Write a report including illustrations detailing where all the tools, equipment and materials are kept in the workshop. Emphasise in your report the health and safety aspects behind the correct storage of tools and materials. On completion, show the report to your tutor for assessment.

- training of personnel to use equipment safely and following manufacturers' or suppliers' instructions;
- accidents involving work equipment happens all the time – many serious, some fatal.

Workplace Transport

Every year, about 70 people are killed and approximately 2500 seriously injured in accidents involving vehicles at the workplace. Being struck or run over by moving vehicles, falling from vehicles or vehicles overturning are the most common causes. There are many different types of vehicles on a construction site at any one time, from cars to much larger vehicles (Figure 1.14).

Often, there is significantly more danger from vehicles in the workplace than on the public highway, since the operating conditions are different.

Slipping and Tripping

The most common cause of injuries at work is the slip or trip. Resulting falls can be serious. They happen in all kinds of business, with the construction industry reporting higher-than-average numbers. These cost employers many millions of pounds a year in lost production and other expenses.

Hazardous Substances

Thousands of people are exposed to all kinds of hazardous substances at work. These can include chemicals that people make or work with directly and dust, fumes and bacteria, all of which can be present in the workplace. Exposure can happen by breathing them in, contact with the skin, splashing them into the eyes or swallowing them. If exposure is not prevented or properly controlled, it can cause serious illness, including cancer and dermatitis, trigger asthma and sometimes even cause death.

Storage of Combustibles and Chemicals on Site

Liquefied petroleum gas (LPG), petrol, cellulose thinners, methylated spirits, and white spirit are all highly flammable liquids and require special storage to ensure they do not risk injury to operatives.

Small supplies of flammable substances should be kept in clearly marked containers with securely fastened caps or lids in a well-ventilated store built with brick walls, concrete floor and non-combustible roof. The store should be sited well away from other buildings and marked with a prominent notice stating that the contents are highly inflammable.

Figure 1.14 Workplace transport: a site stacker truck.

Remember Keep a check on your workplace safety.

Note

Safety programmes should develop a permanent safety consciousness in learners and workers by promoting the idea of doing things the safe way.

Figure 1.15 LPG storage.

Liquid Petroleum Gas (LPG)

As well as the above, there are also specific storage regulations for liquid petroleum gas. LPG must be stored in the open and usually in a locked cage. It should be stored off the floor and protected from direct sunlight and frost or snow (Figure 1.15).

The storage of LPG is covered by the Highly Flammable Liquids and Liquefied Petroleum Gases Regulations.

Chemicals

Chemicals such as brick cleaners or certain types of adhesives or solvents can be classified as dangerous chemicals and must be stored in a locked area to prevent abuse or cross-contamination. Always check the COSHH before handling and storing chemicals.

Health and Hygiene

Everyone who works on site must have access to adequate toilet and washing facilities, a place for warming up and eating their food and somewhere for storing clothing. Toilets need to be easily accessible from where the work is being done. Wash hand basins should be close to toilets. Washing facilities need to be near rest rooms so that people can wash before eating.

In the building industry, you will be exposed to substances or situations that may be harmful to your health. Some of these health risks may not be noticeable straight away and it may take years for symptoms to be recognised.

Sanitary Conveniences

The number of toilets required will depend on the number of people working on the site. A wash basin with water, soap and towels or dryers should be close to the toilets if the toilets are not near the other washing facilities provided on the site.

Washing Facilities

On all building sites, basins should be provided large enough to allow people to wash their faces, hands and forearms. All basins should have a supply of clean hot and cold, or warm, water. Water supplied from a tank may be used. Soap and towels or dryers should also be provided.

Where the work is particularly dirty or workers are exposed to toxic or corrosive substances, showers may be needed.

Drinking Water

There should be a supply of drinking water. It is best if a tap direct from the mains is available. Otherwise, bottles or tanks of water may be used for storage.

Storing and Changing Clothes

There should be arrangements for storing (Figure 1.16):

- clothing not worn on site (e.g. hats and coats)
- protective clothing needed for site work (e.g. wellington boots and overalls).

Separate lockers might be needed, although on smaller sites the site office may be a suitable storage area, provided it is kept secure. Where there is a risk of protective site clothing contaminating everyday clothing, items should be stored separately.

There should also be somewhere to dry wet clothing.

Rest Facilities

Facilities for taking breaks and meals should be available. The facilities should provide shelter from the wind and rain and be heated as necessary.

The rest facilities should have:

Figure 1.16 Example of a storage facility: a locker.

- tables and chairs;
- a kettle or urn for boiling water;
- a means for heating food (e.g. a gas/electric ring/microwave oven).

For small sites, rest facilities can often be provided within the site office, or site hut, especially where this is one of the common portable units.

Noise Safety

Noise is the sound made by pressure changes in the air and picked up by your ear. Loud noise can annoy people. More importantly, it can damage your hearing. But very soft noise can be difficult to hear.

People who are exposed to high noise levels, even for a short time, may experience temporary hearing loss. If they are exposed to noise for a long time, they can suffer serious, permanent hearing loss. Sufferers do not often realise that their hearing is being damaged until other people notice and make them aware of it.

The damage happens when pressure changes in the air affect the inner ear. This is the part of the ear that allows you to hear. You will find that loud noise over a short period of time can cause temporary hearing loss and a 'buzzing' in your ears.

At work, noise can stop you concentrating. It distracts you and may make you unsafe. There is legislation in place to help protect your hearing throughout your lifetime.

Noise is measured in decibels (dB).

As a guide:

- If you have to raise your voice to speak to someone 2m away, noise levels are about 85dB.
- If you have to shout to speak to someone who is 1m away, noise levels are about 90dB.

Identifying 'ear protection zones' and putting up signs where noise is at or above 90dB can control the effects of noise.

Noise at Work Regulations

Noise levels are measured with sound level meters. They have up to four scales, A to D, which give readings in decibels (dB). The most common scale for construction work and for legal purposes is the A scale.

The regulations identify time limits for exposure to various sound levels and set out three action levels:

- First level 80dB (A) scale: Employee is provided, at their request, with suitable and efficient personal ear protectors.

- Second level 85 dB (A) scale: Employee is provided with suitable personal ear protectors, which must be worn.
- Peak level 140 dB (A) scale: Employee must wear the personal protective equipment (PPE) provided as noise at this level will cause permanent damage to hearing.

Noise assessment should be carried out by a competent person.

Wearing Ear Protection

You should wear ear protection when the sound level is between the 80 dB and 85 dB action levels. You must wear it above 85 dB.

Without protection, there is a risk of damage to your hearing. Remember that, over time, this damage can result in permanent hearing loss. Ear protection cannot repair damage that has already been caused.

Substances Hazardous to Health

Any hazardous substances that are going to be used, or processes which may produce hazardous materials, should be identified. The risks from work which may affect site workers or members of the public should then be assessed. Designers should eliminate hazardous materials from their designs. Where this is not possible, they should specify the least hazardous products that perform satisfactorily.

What is a substance hazardous to health under the COSHH? Under COSHH, there are a range of substances that are regarded as hazardous to heath.

Hazardous substances include:

- substances used directly in work activities (e.g. adhesives, paints, cleaning agents);
- substances generated during work activities (e.g. fumes from soldering and welding);
- naturally occurring substances (e.g. grain dust);
- biological agents such as bacteria and other micro-organisms.

Substances or Mixtures of Substances Classified as Dangerous to Health under the Chemical Regulations 2002

A warning label can identify these substances. The supplier must provide a safety data sheet for them. Many commonly used dangerous substances are listed in the HSE's *Publication Approved Supply List*.

Any kind of dust if its average concentration in the air exceeds the levels specified in COSHH should be controlled.

For the vast majority of commercial chemicals, the presence (or not) of a warning label will indicate whether COSHH is relevant.

Advice and Information

If in doubt, contact your local HSE office (the address is in the phone-book). The staff there can refer you to the appropriate inspector or the environmental health officer at your local authority.

Assessment

Your supervisor will look at the way people are exposed to the hazardous substance in the particular job that is about to be done. He will then decide whether it is likely to harm anyone's health.

Prevention

If harm from the substance is likely, the first step to take is to try to avoid it completely by not using it at all.

Control Exposure

If the substance has to be used because there is no alternative, the next step is to try to control exposure. Some of the ways this could be done include:

- ensuring good ventilation in the workplace by opening doors and windows;
- using as little of the hazardous substance as possible;
- transferring liquids by a pump or siphon rather than by hand;
- using cutting or grinding tools fitted with exhaust ventilation or water suppression to control dust.

Personal Protective Equipment

If, and only if, exposure cannot be adequately controlled by any combination of the measures already mentioned, provide personal protective equipment. This can take the form of:

- respirators, which can protect against dusts, vapours and gases;
- protective clothing, such as overalls, boots and gloves;
- eye protection, such as goggles or face visors.

Select PPE with care. Choose good-quality equipment which is CE-marked. (See section on PPE elsewhere in this chapter.)

Personal Hygiene

Substances can be a hazard to health when they are transferred from workers' hands onto food, and so taken into the body. This can be avoided by good personal hygiene, for example by:

- washing hands and face before eating, drinking and before using the toilet (Figure 1.17);
- eating and drinking only away from the work area.

Words and Meanings

Hazardous – Dangerous or harmful.
Substances – Materials used in the course of the work activity.

Figure 1.17 Handwashing.

Try this Out

During your first week at college, your tutor will inform you of basic hygiene standards and health and safety issues in the workshop.

Carry out a report listing the procedures you have to carry out at the end of every workshop session. These may include:

- putting away all materials, tools and equipment;
- sweeping the floor and disposing of waste matter in an environmentally friendly manner;
- washing your hands and face;
- placing your overalls back in your locker and so on.

In your report include all health and safety aspects of your time in the workshop with the emphasis on why things are done in a certain way.

Make sure as few people as possible are exposed to the substances by excluding people not directly involved in the work from the contaminated area.

Make sure those at risk know the hazards. Provide good washing facilities and somewhere clean to eat meals. Good clean welfare facilities can play an important part in protecting the health of everyone involved in the work.

Drugs and Alcohol

Excessive drinking and the use of drugs by operatives is of serious concern to employers, given the risk of the operative underperforming and putting their own health and safety, and that of others, at risk.

Drug abuse poses a potential threat to the health, well-being and the livelihood of operatives. The consequences are a reduction in perception, concentration and awareness, which can affect the safety and welfare of afflicted persons and those of others. The inability of a person to function competently and with reasonable care is a problem that must be addressed to prevent accidents occurring in the workplace.

Alcohol-related problems can be detrimental to the individual and the smooth running of the site and can result in waste and inefficiency. However, it is well known that such problems can be effectively treated. Source: Construction skills, 2007.

Waste Control

Over 70 million tonnes of waste is produced in the construction industry each year. This amounts to 24 kg per week for every person in the United Kingdom, about four times the rate of household waste production. Government guidance suggests we should follow a prioritising approach to reduce the amount of waste and to re-use and recycle what is produced.

It is important that all waste material is collected and removed from the work area promptly and not allowed to accumulate. When waste has accumulated, it will require removal. The most common method is loading into a waste disposal skip (Figure 1.18).

When cleaning the work area, there may be many materials that can be salvaged and used again. Surplus bricks, blocks and so on should be returned to their original storage place. Materials should be properly cleaned, if necessary, before storing.

When work is being carried out at height, the waste materials need to be removed to the ground level without causing a dust problem. Waste chutes are used, and these can be adapted to suit any height (Figure 1.19). They are utilised to empty waste material into skips often with a large dust sheet attached.

Words and Meanings

Waste – All substances that the holder wishes, or is required, to dispose of in solid, liquid or gaseous forms.

Remember Materials are expensive – do not throw them away.

Figure 1.19 Example of a waste chute.

Figure 1.18 Example of an empty skip.

Using Personal Protective Equipment

Types of PPE used in the workplace including hard hat, face mask, eye shield, breathing apparatus, dust mask, high-visibility (or hi-vis, as they are sometimes known) jackets, steel-toecap boots, ear defenders, gloves, sun protection, barrier cream and clothing.

The use of protective clothing and equipment is not the solution for preventing accidents. It is most important that the primary protection against accidents is to identify possible hazards and take the necessary safety measures to eliminate the hazards.

The wearing of protective clothing should provide a back-up for a safety programme.

Head Protection

Safety helmets or hard hats should be worn by workers employed at any place where they may be exposed to head injury from:

- falling
- falling or flying objects
- striking against objects or structures.

Where necessary to protect the head from possible electrical shocks, protective hats should be insulated or made of insulating materials. See Figure 1.20.

Workers working in the sun in hot weather should wear suitable head covering.

Figure 1.20   Examples of head protection.

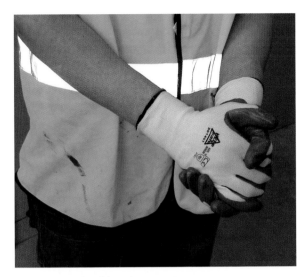

Figure 1.21   Example of hand protection.

Hand Protection

Where necessary, workers should wear suitable gloves or gauntlets (Figure 1.21) and/or be protected with appropriate barrier creams when employed at places where they may be exposed to hand or arm injuries from:

- corrosive or toxic substances
- sharp or rough points, edges or surfaces.

Foot Protection

Workers should wear footwear of an appropriate type when employed at places where they may be exposed to injury from:

- falling objects
- hot, corrosive or poisonous substances
- sharp-edged tools (Figure 1.22)
- nails
- abnormally wet surface
- slippery or ice-covered surfaces.

Eye Protection

Workers should be protected by a screen or wear clear or coloured goggles or other suitable eye protection when employed at places where they may be exposed to eye injury from:

- flying particles (Figure 1.22)
- dangerous substances
- harmful light or other radiation.

Figure 1.22   PPE: Example of eye and foot protection.

Try this Out

Practical Exercise: Cutting Bricks and Blocks

Your tutor has asked you to sort through the PPE he has provided and to choose the following to suit your size: overalls, safety helmet, gloves, safety footwear and goggles.

Adjust the PPE accordingly so that it fits you well. Get your tutor to check that you have selected and adjusted the PPE correctly.

Maintain PPE

After completion of the cutting bricks and blocks exercise, ensure all the PPE you used is cleaned and put away in the proper place.

Get your tutor to check to see if you have carried out this activity to the required industrial standard.

Moving Vehicles

Workers who are regularly exposed to danger from moving vehicles should wear:

- distinguishing clothing, preferably bright yellow or orange in colour; or
- devices of reflecting or otherwise conspicuously visible material (e.g. high-visibility jackets).

Vehicles such as earth-moving equipment, forklifts, ready-mixed concrete delivery trucks (Figure 1.23) and so on should have a distinct warning signal that activates automatically when the vehicle reverses.

Equipment Checks

Frequent safety checks of workshop equipment, building sites, scaffolding and safety equipment such as fire extinguishers and so on should be mandatory in any safety programme.

Storage and Maintenance of Personal Protective Equipment

The correct storage and maintenance of personal protective equipment is important because damaged personal protective equipment will

Figure 1.23 Site vehicle: a ready-mixed concrete delivery truck.

probably not provide you with the necessary protection required against site hazards.

To be effective, all PPE that you use must be manufactured to a known standard, kept in good condition and be suitable for the user. For example, a damaged safety helmet will not offer the required degree of protection as an undamaged one would.

Health and safety law places a legal duty on you not to interfere with or misuse anything provided for your safety. This means looking after and maintaining PPE correctly.

Importance of Personal Protective Equipment

Depending on the type of site situation or workshop, the wearing of personal protective clothing, the use of the correct safety equipment and safe practices are the best ways of avoiding accidents or injuries.

Consequences of Not Using Personal Protective Equipment

The possible health risks of not using PPE in the workplace are:

- catching skin cancers, dermatitis and various infections;
- damage to the eyes, cuts and head injuries;
- leptospirosis and burns;
- hearing damage and respiratory failure.

Training is needed in the use of some PPE. If you are not trained, the PPE will not be effective and you will be exposed to the dangers from which the PPE was supposed to protect you.

Remember It is in your interests to correctly store and maintain PPE and is a legal requirement.

Remember You have a legal responsibility to wear or use the PPE your employer has provided. You cannot decide to opt out.

Fire and Emergency Procedures

Elements Essential to Creating and Sustaining a Fire

The following three elements need to be present for a fire to start. If you remove one element, you will be able to put a fire out.

- Fuel: can be anything that will burn (e.g. wood, furniture, flammable liquid, gas).
- Oxygen: or air in normal circumstances will allow a fire to burn.
- Heat: a minimum temperature is needed but a naked flame, match or spark is sufficient to start a fire, especially if in contact with something flammable.

Emergency Evacuation Procedures

During your first week at college or work, your tutor or supervisor will carry out an emergency evacuation procedure (Figure 1.24). This will consist of:

- sounding the alarm
- leaving the building by the nearest available exit
- reporting to an assembly point.

During the procedure, you must not:

- return to the building until authorised to do so
- use the lifts.

Pay attention to the procedure and make notes in your college workbook, as it will be of use to you at a later stage in the programme.

How Fire Can Spread

To reduce the risk of the spread of fire, site offices, stores and other temporary buildings should be sited at least 6 m away from main buildings. Temporary offices should be built with a reasonable distance between them, preferably up to 6 m. The area around site offices should be kept free of combustible materials; these include grass or weeds. Similar precautions should be taken at the perimeter of the compound, where weeds and grass can accumulate and present a hazard to materials stored close to the fence.

Storage areas must be kept free of surplus packing materials and other debris likely to ignite quickly, which should be put in bins and removed at regular intervals. There are various types of fire extinguishers (see Table 1.1 later in this chapter). All fire extinguishers should be

Figure 1.24   Fire evacuation sign.

located so that they can be reached as quickly as possible in the event of fire. They should be identified by a prominent sign.

The employer should nominate a person to be responsible for fire precautions on site who should be available outside working hours so that the fire service or police can contact him/her if a fire breaks out on site.

Fire Prevention

Many fires can be avoided by carefully planning and controlling work activities. Good housekeeping and site tidiness are important not only to prevent fire but also to ensure that emergency routes do not become obstructed. It helps to make, and adhere, to site rules.

Plan how the site can be kept tidy. In particular, walkways and stairs should be kept free of loose materials. Clear all paper, timber offcuts and other flammable materials from all areas to reduce fire risks.

Types of Fire Extinguisher

There are six different types of extinguishers to use to fight fires: water, carbon dioxide (CO_2), foam, powder, halon and wet chemical (Table 1.1). You will need to choose the right one. Each is designed to put out fires that are caused by specific things. It can be dangerous to use the wrong fire extinguisher.

The standard colour for fire extinguishers is red, with the contents indicated by a contrasting colour band or panel on the extinguisher. There are also pictograms showing what type of fire it can be used on.

Fire Blankets

In addition to the various types of fire extinguisher, there are fire blankets (Figure 1.25). These are fireproof blankets that, when laid on a fire, cut off the oxygen supply to the fire and so stifle the flames. They can be used on all types of fires.

Discovering a Fire

Fires are the worst kind of hazard on a construction site. They do a great deal of damage every year. You should investigate all fires, however small, and report them to your supervisor.

Where there is a fire risk, all necessary precautions must be taken. Everyone on site should be aware of the fire drill procedure.

If you discover a fire:

- Raise the alarm.
- Close doors and windows to prevent the spread.
- Evacuate the area.

You are the Supervisor

Checklist: Fire and emergency procedures

- Is the quantity of flammable materials, liquids and gases kept to a minimum?
- Are they properly stored?

You are the Supervisor

Checklist: Fire and emergency procedures

Are smoking and other ignition sources banned in areas where gases or flammable liquids are stored or used?

Try this Out

Simulated Fire Evacuation Procedure

With your colleagues, plan and carry out a simulated fire evacuation procedure of the brick workshop, which must include ensuring all exits are clear and assembling at the correct fire assembly point.

On completion, make a report of the activity and give it to your tutor for checking against the regulations.

Table 1.1 Types of fire extinguishers

Type	Band colour	Use on	Do not use on/in
Water	Red	Solid fuels (e.g. wood, paper, textiles)	Flammable liquids or live electrical equipment
Foam	Cream	Wood, paper, textiles, flammable liquids	Live electrical equipment
Dry powder	Blue	Wood, paper, textiles, flammable liquids, gaseous fires, live electrical equipment	–
Carbon dioxide	Black	Flammable liquids and live electrical equipment	Confined space
Halon	Green	Flammable and liquefied gases and electrical hazards	–
Wet chemical	Canary yellow	Deep fat fryers	All other types of fire

Water

For use on

A Wood, Paper, Textiles

Do not use on

B Flammable liquids

Live electrical equipment

Dry powder

For use on

A Wood, Paper, Textiles

B Flammable liquids

C Gaseous fires

Live electrical equipment

Foam

For use on

A Wood, Paper, Textiles

B Flammable liquids

Do not use on

Live electrical equipment

CO₂ Carbon dioxide

For use on

B Flammable liquids

Live electrical equipment

Do not use in a confined space

All fire extinguishers are red and each has a different colour band to identify its type.

Figure 1.25 Fire blanket.

- Fight the fire, if you have been trained to do so, but avoid endangering life.
- Fight the fire with an appropriate fire extinguisher, fire blankets, water or sand, but do not put yourself at risk.
- Only persons who are fully trained should carry out the fighting of fires.

In order to save lives and reduce the risk of injury occurring, all site personnel must be informed of the current safety and emergency procedures. All site personnel must be aware of what to do in the event of a fire or accident.

Fire Evacuation Procedures

During induction to any workplace, you will be made aware of the fire procedure as well as where the fire assembly points are and what the alarm sounds like. On hearing the alarm, you must stop what you are doing and make your way to the nearest assembly point.

This is so everyone can be accounted for. If you do not go to the assembly point or if you leave before someone has taken a register, someone may risk their life to go back into the fire to get you.

Note

Halon extinguishers are being phased out for environmental reasons.

Remember Fire extinguishers should only be used if the user is properly trained. Untrained users could make the situation worse.

When you hear the alarm, you should not stop to gather any belongings and you must not run. If you discover a fire, you must only try to fight the fire if it is blocking your exit or if it is small. Only when you have been given the all-clear can you re-enter the site or building.

Signs and Notices

Employers are required to provide safety signs in a variety of situations that do, or may, affect health and safety. There are five types of safety signs in general use. Each of these types has a designated shape and colour to make sure that employees get health and safety information in a simple, bold and standard way, and with little use of words (Figure 1.26).

Building sites have many safety signs. These include general signs at the entrance to a site and many other more specific signs displayed

Figure 1.26 Identification of safety signs.

around the site as appropriate (Figure 1.27). Knowing the standard format for different types of sign will help you to interpret these signs effectively, for example knowing that some signs are warnings and others are prohibitive and that these can be distinguished by shape and colour.

Safety signs on site give vital information to keep you and others safe while you are working. They use a standard colour and shape system to make them instantly recognisable, even from a distance.

To be safe at work you will need to recognise, understand and respond to a lot of different safety signs. The five basic categories of safety signs are listed in Table 1.2.

The standard system of safety signs and colours used on site draws your attention to objects and situations that could affect your health or safety. But there are other types of signs. Supplementary signs are signs with writing on them, rather than just symbols. They can be used on their own or in support of other signs to provide more information. For example, the signs:

PROHIBITION
Stop/must not
Red on white background

WARNING
Risk of danger hazard ahead
Yellow background with black border

MANDATORY
Must obey
Blue background with white symbol

SAFE WAY TO GO
Safe condition
Green background with white symbol

No smoking

EYE WASH BOTTLE

Figure 1.27 Safety signs and signals.

Table 1.2 The five basic categories of safety signs

Type	Shape	Colour	Meaning
Warning	triangular	yellow background with a black border and symbol	Warns of hazards or danger (e.g. 'Caution, there is a risk of an electric shock')
Prohibition	circular	red border and cross bar with a black symbol on a white background	Shows what must not be done (e.g. 'No smoking')
Safe	square or oblong	white symbols on a green background	Indicates or gives information about safety provision (e.g. 'First aid available in site office')
Mandatory	circular	white symbol on a blue background	Shows what must be done (e.g. 'Wear your safety helmet')
Fire safety and equipment	square or rectangular	white symbols on a red background	Gives location of fire information, alarms or equipment (e.g. 'Fire extinguisher located in site office')

Try this Out

Think of some examples of safety signs from any sites you have worked on and then draw the signs using coloured pencils. Check with your tutor to see if you have drawn them correctly.

Note: If you have difficulty recognising colours, speak to your tutor or supervisor.

Fire assembly point

Figure 1.28 Fire safety signs.

All drivers and visitors to report to site office

can be on a white background, whereas:

Hand protection must be worn

Remember Always look out for and obey safety signs – they are there for your protection.

can be the colour of the sign it is supporting, in this case blue.

Fire safety signs are of equal importance and are there for the safety of both workers on and visitors to a site. Figure 1.28 shows two of the most common fire safety signs.

Health and Safety Information

There are various sources of health and safety information. For more details on all aspects of health and safety, visit the following institutions' websites.

The Health and Safety Executive

The HSE website contains information about the objectives of HSE, how to contact HSE, how to complain, recent press releases and research and current initiatives. Information about risks at work and information about different workplaces is also available. HSE priced publications are also available from bookshops and free leaflets can be downloaded from the HSE website.

Website: www.hse.gov.uk

ConstructionSkills

ConstructionSkills is the sector skills council for construction. It represents every part of the construction industry, from architects to bricklayers, in every part of the United Kingdom. And it covers every part of the skills agenda – from grants to college places.

Website: www.cskills.org

Royal Society for the Prevention of Accidents (RoSPA)

The Royal Society for the Prevention of Accidents is a registered charity whose mission is to save lives and reduce injuries.

Website: www.forms.rospa.com

British Safety Council

The British Safety Council is one of the world's leading occupational health, safety and environmental organisations. Founded in 1957, it now has a turnover of more than £9 million. Its mission is to support a healthier, safer and more sustainable society.

Website: www.britsafe.org

CHAPTER 2

The Handling and Storage of Materials

THIS CHAPTER RELATES TO UNIT CC1001K AND UNIT 1001S.

Handling Materials

Lifting heavy or awkward objects, such as bags of cement and plaster, can cause injury if not performed correctly. Incorrect lifting techniques can put stress on the lower back. After years of bad lifting, the discs between the various vertebrae become disjointed and are prone to slipping. By using the 'kinetic method' of manual handling, injuries to the back can be avoided.

The kinetic method is based on two principles:

- fully employing the strong leg muscles for lifting, rather than the weaker muscles of the back;
- using the momentum of the weight of the body to begin horizontal movement.

These two motions are combined in smooth continuous movements by correct positioning of the feet, maintaining a straight back and flexing and extending the knees (Figure 2.1).

In practice, this requires the correct positioning of the feet, a straight back, arms close to the body when lifting or carrying, the correct hold, keeping the chin tucked in and using the body weight.

The correct procedures for certain types of handling are described in the next pages.

Words and Meanings

Kinetic Lifting – **Term used to describe the correct method for lifting.**

Manual Lifting and Carrying

The Health and Safety Executive (HSE) outlines important handling points, using a basic lifting operation as an example.

Figure 2.1   Correct lifting techniques.

Positioning the feet

Figure 2.2   Positioning the feet.

> **Note**
>
> It is recommended that the feet be placed about 50 cm apart. This distance is suitable for a person having a height of about 175 cm.

> **Note**
>
> Do not attempt to lift or carry any load exceeding 25 kg alone.

Stop and Think

Plan the lift. Where is the load to be placed? Use appropriate handling aids if possible. Do you need help with the load? Remove obstructions such as discarded wrapping materials. For a long lift, such as floor to shoulder height, consider resting the load midway on a table or bench to change grip.

Position the Feet

Injury to the back muscles is often caused by loss of balance owing to working with the feet too close together when lifting, pushing or pulling. The feet should be positioned with one placed in the proposed direction of movement and the other where it can push the body. Figure 2.2 shows the correct positioning of the feet when moving a load in a sideways direction to avoid twisting the trunk.

During manual handling, at no time should the feet be close together on the ground.

Keep the arm and body as straight as possible when lifting heavy objects. Place your feet close to the object.

Bend your knees, squat and keep your back as straight as possible. Lift with the legs – not with the back. If the object is too heavy or too bulky, get help. Figure 2.3 shows the correct posture to adopt when lifting heavy objects.

Figure 2.3 Good posture.

Figure 2.4 Weight on thighs.

Arms Close to Body

When lifting and carrying loads, the arms should be kept close to the body and as straight as possible. This will avoid unnecessary strain on the upper arm muscles and the chest. Also, if the load has to be carried a long distance, the weight can rest on the thighs (Figure 2.4).

Correct Hold

An insecure grip may be due to taking the load on the fingertips caused by a badly designed handle. Greasy surfaces also often prevent a secure hold. Whenever possible, use a full palm grip. This gives a stronger hold and decreases the possibility of the load slipping.

Loads from the Floor

When lifting a load, the chin should be tucked in. This will automatically raise the chest and thus prepares the shoulders for more efficient arm movements.

Now, applying the principles described, a load should be lifted from the floor as follows. Bend the knees, keeping the back straight and the chin tucked in, grasp the load firmly, keeping the arms close to the body, and then straighten up by flexing the knees as illustrated in Figure 2.1.

Note

Always wear protective gloves to avoid cuts, abrasions and splinters.

Figure 2.5 Put down and adjust.

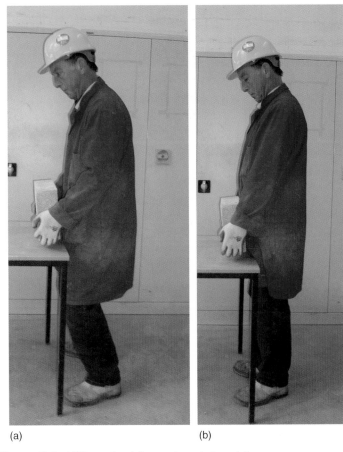

(a)   (b)

Figures 2.6 Lifting a load from a bench 1 and 2.

Put Down and Adjust

If precise positioning of the load is necessary, put it down first and then slide it into the desired position (Figure 2.5).

Lifting A Load from A Bench

Keep your back straight, extend your arms in front of you and bend your knees slightly, until you can grasp the load firmly. Pull the load towards you, straighten up and lean back slightly (Figure 2.6).

Lifting and Carrying Long Loads

In general, a load longer than 6 m requires more than one person to lift and carry it.

When being lifted by one person (Figure 2.7), one end is raised above shoulder level. The operative then walks forward, moving his or

Figure 2.7 Lifting long loads.

her hands along the length until the point of balance is reached. The load is then balanced.

Handling Sheet Materials

Large sheets of material are awkward shapes to pick up. By using a hook device, with a long handle, a large sheet of material can be lifted and carried quite easily.

Handling a Drum

The correct position for the feet and hands when preparing to upend a large drum is as follows. The arm is lifted by pushing with the back foot and extending the legs and at the same time positioning the hands on opposite sides of the rim. To allow the drum to settle on its base, the bodyweight is used as a counterbalance by straightening the back leg.

Alternatively, sheet materials and other large items can be moved by a walking process. This is done by lifting one side and swivelling the item on the opposite corner, repeating the process on alternate corners. Care should be taken to avoid damaging corners when walking items.

Handling Bricks and Blocks

Before brickwork can begin, the working area must be loaded out with bricks and blocks (Figure 2.8).

The objective is to:

- stack bricks and blocks safely and within easy reach of the bricklayer;
- space boards at the corners of the work and not more than 3 m apart along the wall length.

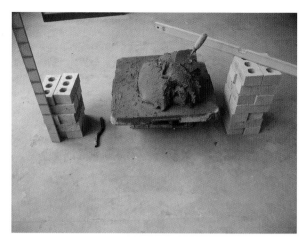

Figure 2.8   Example of a typical workstation.

Brick Tongs

Bricks can be carried by hand from the stack to where they are required, but this is laborious and slow. Brick tongs, which can pick up six bricks on edge, save time and effort.

Brick tongs are manufactured or produced from metal. They are designed to clasp or clamp the load of bricks when they are lifted. There are many different designs of tongs available.

Wheelbarrows, Trolleys and Forklifts

Loose materials such as aggregates, bricks, blocks and bags of cement may be moved by using a wheelbarrow or trolley. Building sites can form rough terrain, and it may be hard work pushing loaded barrows over such rough ground. In these circumstances, barrow runs may be used. These are boards laid down over the ground. Always load the barrow evenly with most of the weight above the wheel (Figure 2.9).

Trolleys are best used for moving items on hard surfaces, such as in warehouses or workshops, since the relatively small wheels will sink into soft surfaces. It is better to take several smaller loads than to risk injury by overloading wheelbarrows or trolleys.

The most common method of moving bricks on larger sites is by an all-terrain forklift truck.

Team Lifting

If an object has been assessed as being too heavy or awkward for one person to lift, two people should lift it. When team lifting, those lifting should be approximately the same height and build as each other. The

Figure 2.9 Using a wheel barrow.

Figure 2.10 Personal protective equipment suitable for lifting and carrying.

effort should be the same for each person, and only one person should be responsible for giving instructions.

These instructions should be clearly given, using a recognised call such as, 'Lift after three: one, two, three, lift.' Prior to lifting, any objects that are in the immediate area should be removed. It is important that you wear suitable personal protective clothing such as boots, gloves and overalls (Figure 2.10).

Examining Loads

Always examine the load in order to ascertain whether you can lift it; if you feel that this is not possible, obtain help. Before carrying anything, always check the route you will use. Clear the route of any obstructions. Ensure that the place you are taking the object to is clear of obstruction, too.

Bagged Materials

The carrying of bagged materials (Figure 2.11) such as cement and plaster can be very tiring, as the bags are awkward to lift. The easiest way to carry them is to place the bag on the shoulder. This is made easier if you support the shoulder. It may make it easier still if you support your shoulder by putting your hand on your hip and hold the bag with your other hand.

Figure 2.11   Carrying bagged materials.

Storing Materials

Storing Bars and Tubes

Often it is necessary to store metal bars, tubes and similar materials for short periods or for maintaining stock. The type of storage rack illustrated in Figure 2.12 is frequently used in workshops and allows for the separation of materials, which enables quick identification of sizes and types. This type of rack can also be used for storing scaffolding tubes or pipes.

Pipes and other types of cylindrical materials are often brought to a site for immediate use. These materials can be stored on the floor.

Use a wedge or some other blocking device to prevent the stacked pipes from moving or rolling during stacking or removal.

Figure 2.12 Storing tubes.

Remove large diameter pipes from a stack by pulling them out from the ends. Do not remove by lifting from the sides.

Reinforcing Steel

Reinforcing steel should be stored and grouped by diameter in order to facilitate identity and handling.

Bagged Cement

Bags of cement, lime or similar materials should never be stacked more than ten bags high, unless supported by walls. Bags should be stored in a dry place. When stacking bags near the mixing site, place sacks on boards or concrete blocks to avoid wetting from surface water. Canvas or plastic covers should be available to protect against rain or dampness (Figure 2.13).

Bricks and Blocks

Bricks and blocks should be stored on a level base such as pallets, planking or concrete. Do not stack materials more than 2m high.

Note

To avoid accidents, do not stand or walk on stacked pipes when placing or removing pieces.

Figure 2.13   Storage of cement.

Manual Handling Legislation

In addition to the responsibilities of the employer and employee as set out in the Health and Safety at Work Act (see Chapter 1 for an in-depth look at relevant legislation), there are regulations relating to manual handling.

The Manual Handling Operations Regulations 1992

The Manual Handling Operations Regulations 1992 place a requirement on the employer to deal with risks to the safety and health of employees who have to carry out manual handling at work.

These regulations cover all work activities in which a person does the lifting instead of a machine.

The Lifting Operations and Lifting Equipment Regulations 1998

These regulations cover the operation of all lifting equipment, including those which lift people. Information on the regulations can be found in the Approved Code of Practice and guidance on the safe use of lifting equipment.

These regulations cover all work activities in which a machine does the lifting instead of a person.

Taking Delivery of Materials

Your company's materials and equipment may come from an outside supplier, a plant hire firm or your own firm's yard. Whoever supplies them, you must be satisfied that they are exactly what was ordered and that they are in a good condition.

Before unloading a delivery vehicle (Figure 2.14) or receiving a plant item, you need to check that:

- the delivery has come to the correct location;
- the delivery note details match those on the order;
- there are no signs of physical damage;
- unloading can take place safely.

You or your supervisor should ensure that:

- operatives unloading are trained and supervised;
- safety provisions are observed and any supplier unloading instructions are met;
- materials can be checked for damage caused during unloading;
- simple quantity and quality checks are made;
- you have all the necessary information before you sign the delivery note.

At this stage, you may be aware of discrepancies, such as:

Figure 2.14   Unloading a vehicle.

- incorrect quantity or specification;
- damage caused in transit;
- poor quality.

If you become aware of any of these, record full details on all copies of the delivery note. Inform your supervisor if the discrepancies are serious. When you have recorded the delivery on your materials or plant schedule, send the signed delivery note to the appropriate office or individual without delay.

Information

For more information about the manual handling and storage of materials, visit the HSE website (www.hse.gov.uk) and download free leaflets.

Quick Quiz

1. The kinetic method of lifting is based on two principles. What are they?

2. How many bricks on edge can you pick up with a pair of brick tongs?

3. Describe the storage conditions for cement and lime.

4. What is the maximum height for storing materials?

5. What actions would you take if you noticed a discrepancy in a delivery of materials?

CHAPTER 3

Working at Height

THIS CHAPTER RELATES TO UNIT CC1001K AND UNIT CC1001S.

The Dangers of Working at Height

The use of any form of scaffolding can be dangerous and your chances of being involved in an accident are far greater if you are working on scaffolding. With this in mind, it is worth considering the following points concerning health and safety.

The Health and Safety at Work Act 1974 (HASWA) makes it the duty of everyone at work, both employers and employees, to set up and maintain a safe working environment. Under the HASWA, employees and employers have responsibilities, which if ignored can result in criminal prosecution and, if warranted, a custodial sentence.

The Work at Height Regulations 2005

The Work at Height Regulations 2005 came into effect on 6 April 2005. The regulations apply to work at height where there is a risk of a fall liable to cause personal injury.

The regulations place duties on employers, employees, the self-employed and any person who controls the work of others.

As part of the regulations, 'duty holders' must ensure:

- all work at height is properly planned and organised;
- those involved in work at height are competent;
- the risks from work at height are assessed and that appropriate work equipment is selected and used;
- the risks from fragile surfaces are properly controlled;
- equipment for work at height is properly inspected and maintained.

There is a simple hierarchy for managing and selecting equipment for work at height.

Duty holders must:

- avoid work at height where they can;
- use work equipment or other measures to prevent falls where they cannot avoid working at height.

Falls from Height

Construction workers often work at height (Figure 3.1). In 2003/04, 67 people died and nearly 4000 suffered a serious injury as a result of a fall from height in the workplace.

Falls from height are the most common cause of fatal injury and the second most common cause of major injury to employees, accounting for around 15% of all such injuries. All industries are exposed to the risks presented by this hazard, although the level of incidence varies considerably.

As a result, falls from height are a key priority in the Health and Safety Commission Injury Reduction Programme. The programme was set up in 2008. Its objective was to reduce injury rates by 10% within the first two years of its life.

Experience shows that falls from height usually occur as a result of poor management control rather than because of equipment failure. Common factors include:

- failure to recognise a problem;
- failure to provide safe systems of work;
- failure to ensure that safe systems of work are followed;
- inadequate information, instruction, training or supervision provided;
- failure to use appropriate equipment;
- failure to provide safe plant or equipment.

The Health and Safety Executive (HSE) recommends that employers:

Figure 3.1   Working at height.

- follow good practice for work at height (if they do this they should already be doing enough to comply with the regulations);
- follow the risk assessments that have been carried out for work at height activities and make sure all work at height is planned, organised and carried out by competent persons;
- take steps to avoid, prevent or reduce risk;
- choose the right work equipment and select collective measures to prevent falls, such as guard-rails and working platforms.

Words and Meanings

Duty holder – The person responsible for work at height.

Types of Fall

In 2004–2009, the construction sector accounted for 40% of falls from ladders in the United Kingdom (total number: 1203).

Of these, key categories of fall were:

- falls from ladders: 457
- falls from scaffold: 171
- falls from floors, pavements and roads: 114
- falls from vehicles, plant and earth-moving equipment: 89
- falls from building materials such as bricks, tiles and beams: 34
- falls from surfaces and structures below ground level: 11
- falls from other means, than in key categories: 327

Fall Protection

Falls from the leading edge of roof work need to be prevented. Leading edges are created as new roof sheets are laid, or old ones are removed. Falls from these edges should be prevented, as should falls from roof edges and through fragile materials. All roof edges from which people are liable to fall while work is in progress should be protected.

Barriers

Suitable precautions should be taken to prevent falls (Figure 3.2). Guard-rails, toe-boards and other similar barriers should be provided whenever someone could fall 2 m or more and should always be securely fixed.

Barriers should be strong and rigid enough to prevent people from falling and be able to withstand other loads likely to be placed on them.

Figure 3.2 Example of a barrier.

Figure 3.3 Harness and lanyard.

Figure 3.4 Safety netting.

For example, guard-rails fitted with brick guards need to be capable of supporting the weight of stacks of bricks that could fall against them.

Remember, protection is also required at edges of excavations and where people can fall into water.

Harness and Lanyard

Providing a safe place of work and system of work to prevent falls should always be the first consideration. However, there may be circumstances in which it is not practicable for all or any of the requirements for guard-rails and so on to be provided. Where people may still approach an open edge from which they are liable to fall, other forms of protection will be needed. In some cases a suitably attached harness and temporary horizontal lifeline could allow safe working (Figure 3.3).

A harness will not prevent a fall; it can only minimise the risk of injury if there is a fall. The person who falls may be injured by the impact load to the body when the line goes taut or when they strike against parts of the structure during the fall.

Safety Netting

Work at the leading edge requires careful planning to develop a safe system of work. Nets (Figure 3.4) are the preferred method for reducing the risk of injury from falls at the leading edge, as they provide protection to everyone on the roof. The erection of nets should be carried out by trained riggers.

Airbags

Airbag safety systems are a form of soft fall-arrest and comprise inter-linked modular air mattresses. The modules are connected by flexible couplings and are inflated by a pump-driven fan. As the individual airbags fill with low-pressure air, they expand together to form a continuous protective safety surface, giving a cushioned fall, thereby preventing serious injury.

Falling Materials

The risk of falling materials causing injury should be minimised by keeping platforms clear of loose materials. In addition, provide a way of preventing materials or other objects rolling, or being kicked, off the edges of platforms.

This may be done with toe-boards, solid barriers, brick guards or similar at open edges. If the scaffold is erected in a public place, nets, fans or covered walkways may be needed to give extra protection for people who may be passing below.

Brick guards should be positioned so that they are prevented from moving outwards by the toe-board (Figure 3.5). High-visibility barrier netting is not suitable for use as a fall-protection device.

Figure 3.5 Example of a toe-board and brick guard.

Safety Organisation

Construction sites should have a safety officer employed by the contractor to ensure that all work areas are safe. On building sites where there is no safety officer, the foreman or supervisor is usually responsible for the implementation of site safety.

Try this Out

Prepare a list of safety considerations that site personnel need to observe when erecting, taking down and dismantling access equipment.

Present the list to your tutor and ask him or her to compare it with the official procedures in order to check the correctness of your findings.

Basic Working Platforms

A working platform or form of access is essential to a great deal of construction activities, particularly bricklaying. There are a number of systems available, some of which are described below.

There are four main types of access equipment used in relation to brickwork:

- mobile and static towers
- trestle scaffolds
- independent scaffolds
- putlog scaffolds.

Mobile and Static Towers

Mobile and static towers are quickly assembled staging, usually with wheels for ease of movement (Figure 3.6).

Trestle Scaffolds

A trestle scaffold (Figure 3.7) is usually a basic working platform supported on A frames, bandstands or similar type folding supports.

Trestles are only intended for work of a short duration. A more substantial scaffold would be required if work were to go on for any length of time.

Independent Scaffolds

Independent scaffolds (Figure 3.8) usually provide a working platform around an existing building. They require bracing and ties for stability. The erection sequence requires two rows of standards.

Figure 3.6   Example of a mobile tower.

Figure 3.7 Example of a trestle.

Figure 3.8 Independent scaffolding.

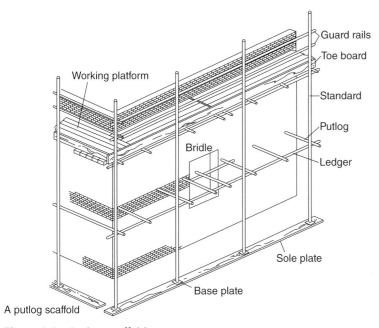

A putlog scaffold

Figure 3.9 Putlog scaffolding.

Putlog Scaffolds

A putlog scaffold, or bricklayer's scaffold, is normally used on the construction of new brickwork (Figure 3.9).

A putlog scaffold is built into walls as brickwork progresses. The scaffold is dependent upon the building for support and stability.

Try this Out

Your tutor has asked you to build a wall in the brick workshop with a colleague that will require the use of an access platform. Plan the work with your colleague, ensuring you avoid risk or harm to yourselves and others.

On completion of the plan, show it to your tutor to see if it meets the specifications.

Component Parts of Independent and Putlog Scaffolds

Scaffolding is made up of several components (Figure 3.10):

- Standard – The vertical tubes which carry the load to the ground. Each standard should sit on a base plate which spreads the load, thus ensuring that it does not sink into the ground.
- Base Plate – Usually made from steel, they have a central spigot which locates the tube. Sometimes the base plate is nailed

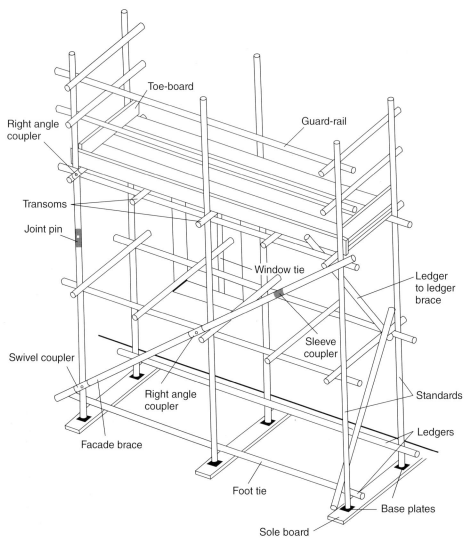

Figure 3.10 Scaffold components.

or screwed to a sole board in order to prevent sideways movement.

- Sole Board – Sole boards or sole plates are necessary, particularly on soft ground, to spread the load over a larger area.
- Toe-board – A board fixed on edge to prevent tools, materials or feet slipping off the platform.
- Guard-rail – A tube fixed to the standards to prevent workers falling off platforms.
- Ledgers – The horizontal tubes which connect and support the standards and act as the support for transoms.
- Lift – The distance between ledgers.
- Main Transoms – Tubes positioned at right angles across the ledgers, next to each pair of standards or connected to each pair of standards. Their function is to hold standards in place, help make the scaffold more rigid and act as scaffold board supports.
- Intermediate Transoms – Tubes positioned across the ledgers between the main transoms to act as scaffold board supports.
- Facade Brace – Tubes fixed to the face of the scaffold to stop the scaffold moving. They should run from the base to the full height of the scaffold at an angle of between 35 and 55 degrees and be fitted at the base and at every lift level, either to the standards or the ends of transoms.

Using Basic Working Platforms

Try this Out

Erect Working Platforms

Your tutor has asked you, under his supervision, to erect the following basic working platforms:

- bricklayer's trestle.

Before carrying out the activities, ensure you have carried out a risk assessment and are wearing the correct personal protective equipment (PPE), are aware of all health and safety issues and only carry out the activity under the supervision of your tutor. On completion ask your tutor to assess your work against the industrial standards.

Using PPE

Before carrying out the erection of the basic working platform activity, draw up a list of PPE you may need. This may include safety helmet, safety boots, eye and ear protection, gloves and, if you are working in bright sunshine, face protection and sun cream.

On completion of the list, show it to your tutor for checking against the industrial standards.

Stepladders

Stepladders are one of the most common forms of access used within the construction industry (Figure 3.11). They are often taken for granted and not as frequently checked for faults as they should be.

When using stepladders, always observe the following procedures:

- Never use stepladders on a working platform (e.g. a tower scaffold).
- Never use a stepladder in the closed position as you would a ladder.
- Never paint a stepladder as this could hide any potential defects.
- Always store under cover after use.

Those stepladders manufactured to the British Standard and which display the Kitemark are the most preferable to use (Figure 3.12).

Ladders and Extension Ladders

Ladders, and extension ladders (Figure 3.13), are best used as a means of getting to a workplace. They should only be used as a workplace for short-term work. They are only suitable for light work.

Figure 3.11 Example of a stepladder.

BSI Kitemark
certification symbol

Figure 3.12 BSI Kitemark certification symbol.

Figure 3.13 Extension ladder.

If ladders are to be used, make sure:

- the work only requires one hand to be used;
- the work can be reached without stretching;
- the ladder can be fixed to prevent slipping;
- a good handhold is available.

In order to use a ladder safely, the person should be able to reach the work from a position 1m below the top of the ladder.

Many accidents result from using ladders for a job when a tower scaffold or mobile access platform would have been safer and more efficient.

For safe use, the ladder needs to be strong enough for the job and in good condition.

Component Parts

- Stiles – Ladder stiles are usually made from Douglas fir, Whitewood, redwood or hemlock. The stiles of pole ladders are made from whitewood.
- Rungs – Ladder rungs are round or rectangular, made from oak, ash or hickory.
- Ties – These are steel rods fitted below the second rung from each end, and at not less than 9-rung intervals. Ties can also be fitted under every rung.
- Reinforcing Wires – Reinforcing wires give ladders extra strength. Galvanised wire or steel cable is fitted and secured, under tension, into grooves in the stiles.
- Ropes – Ropes are made from hemp sash cord or a material of equal strength. Ropes of man-made fibre must provide the operative with an adequate handgrip.
- Guide Brackets – These are fitted at the top of the lower sections to keep the sections together.
- Latching Brackets – These are fitted to the bottom of extension sections to hook over a rung of the section below.
- Pulley Wheels – These guide and facilitate the smooth running of the ropes of rope-operated ladders.

Trestles

Trestles are used in place of the more expensive scaffolding. They come in various types and sizes, but all scaffolding regulations still apply. However, as a rule trestles should only be used for light work and where there is little risk of injury.

Figure 3.14 Tower scaffold.

The component parts of trestles are as follows:

- transom
- adjusting pin
- adjustable legs.

Adjustable steel trestles or bandstands, as illustrated in Figure 3.7, are preferred by bricklayers for small works for the following reasons:

They can take four scaffold boards to form a suitable working platform. They are stronger than other types of trestle and can therefore take loads such as bricks and blocks.

Ground preparation is essential if these types of trestle are to be used on rough or uneven ground. Trestles should always be placed on flat scaffold boards once the ground has been levelled and compacted.

Tower Scaffolding

Mobile tower scaffolds (Figure 3.14) are used extensively throughout the construction industry. They can be constructed from scaffold tubes and couplers or, more usually, from a proprietary system of purpose-made interlocking components.

Proprietary system mobile towers are the type usually found on construction sites and are usually made from the following materials:

- steel
- aluminium
- glass fibre.

Before the Tower Is Used

The tower must be vertical – square – and all horizontal braces and platforms level.

- Outriggers – correctly positioned and secure and in good contact with the ground. Base plate or castor wheels in full contact with the ground. All spigot joints must be fully home and secure.
- All bracing members must be located exactly as per the instruction manual.
- Guard-rails and toe-boards must be in position and secure.
- All access stairways or ladders must be firmly located.

Component Parts

Scaffolding has many component parts (Figure 3.15), including:

- Outriggers – Outriggers are provided to give extra stability at greater height.

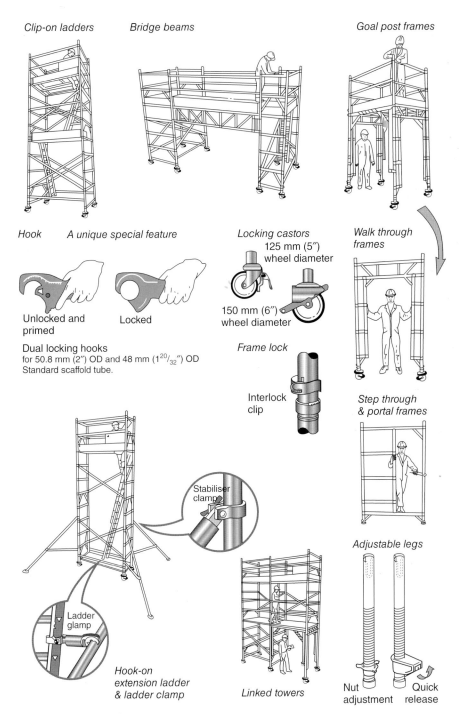

Figure 3.15 Component parts of a tower.

You are the Supervisor

Checklist – Ladders

Ladders and stepladders are the last resort. Can you buy or hire some alternative equipment that would provide a safer means of access?

You are the Supervisor

Checklist – Scaffolds

Is the scaffold strong enough to carry the weight of materials stored on it and are these evenly distributed? Does a competent person inspect the scaffold or proprietary tower scaffold regularly, e.g. at least once a week, and always after it has been altered, damaged and following bad weather?

- Ladders – Internal ladder rungs allow safe access to the working platform.
- Platforms – Working platforms are usually constructed out of box section for minimal deflection.
- Adjustable Legs – Legs with adjustable nuts allow different heights to be reached and allow for the correct levelling of the tower.
- Wheels – Allows the tower to be moved and then locked into place.
- Toe-boards – Toe-board systems are designed to be quick and easy to assemble.
- Guard-rails – Double handrails are usually provided as standard for added safety at all platform levels.
- Braces – Continuous bracing is usually standard, to give the tower added strength.

Try this Out

Carry out research and list the sequence to follow when taking down a ladder from scaffolding or access platforms.

Present your findings to your tutor and check to see if they match official guidelines.

Information

For more information about working at height, obtain free leaflets from the HSE website (www.hse.gov.uk).

1. In 2003/04 how many people died from falls from height?

2. Why do falls from height usually occur?

3. How is the risk of falling material minimised?

4. List five components of a working platform.

5. Which step ladders are the more preferable to use?

CHAPTER 4

Communication and Site Documentation

THIS CHAPTER RELATES TO UNIT CC1002K AND UNIT CC1002S.

Methods of Communication

There are three main methods of using information and communicating effectively: direct speech (talking), written communication (letters, emails etc.) and pictorial (pictures and symbols).

Direct Speech: People Talking and Listening

We probably spend more time talking and listening to people than using any other communication method. If it is done effectively, it is generally the most successful method to use. It gives you direct, personal, contact. You can raise questions, answer them, discuss and settle misunderstandings as they arise.

Some regular communications that rely on people talking and listening are as follows:

- site meetings
- teaching and learning sessions
- team and class briefings
- conversations
- interviews
- telephone and so on.

Written and Printed: People Writing and Reading

Many supervisors, tutors and learners complain about the quantities of paperwork they have to handle. However, as long as it is relevant to you and your organisation, written communication (Figure 4.1) does have advantages over other general communication methods.

Brickwork and Blockwork by Joseph Durkin. © 2011 Joseph Durkin. Published 2011 by Blackwell Publishing Ltd

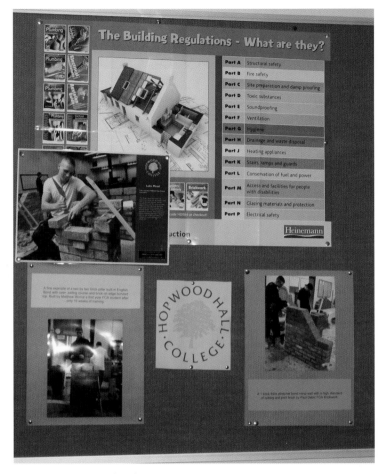

Figure 4.1 Example of written communication.

Written and printed communication on site and at college will include:

- specifications
- schedules
- instructions
- operating manuals
- notices
- posters
- memos
- reports
- letters
- forms
- agendas and minutes
- and many others.

Visual and Graphical: People Using Pictures and Symbols

The construction industry makes great use of visual and graphical means of communication such as working drawings, plans, diagrams, illustrations, bar charts (Figure 4.2), tables, graphs, photographs, models and many more.

Conveying a message largely through some kind of picture or symbol is useful because:

- it can show techniques, methods and whole processes clearly
- it's often easier to understand than the same information given in written or spoken words
- you can use it to support writing or speech (visual aids, illustrations)

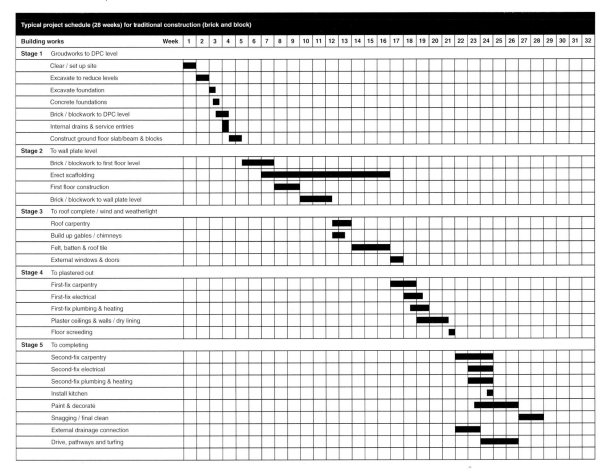

Figure 4.2 Example of a typical bar chart.

- it can aid the understanding of those with language or reading difficulties.

Selecting the most suitable method for your purposes is a central part of planning your communication. The layout/design of a notice explaining a company's health and safety policy

It is the policy of this company (Joseph Ellam Construction) that all site personnel must wear or use the correct personal protective equipment at all times whilst on our sites.

will be different from that of a list of site regulations:

- All health and safety guides must be read and adhered to.
- Before, during and after carrying out any work, the workplace must be clean and tidy to prevent tripping and falling.
- Organise the work and set up tools and equipment.
- Check all equipment is in good working order.
- Always wear the appropriate personal protective equipment.
- Be aware of emergency procedures in the event of an accident.
- Be aware of emergency procedures in the event of a fire.

Why is Effective Communication so Important?

We use communication to create and maintain good working relationships with our supervisor or tutor, our colleagues or anyone else we work with. Effective communication helps us to work together and get the job done. Supervisors and tutors may have authority over members of their working teams or classes. But they cannot exercise that authority successfully if they are not able to communicate effectively … just as students will find that the best way to deal with any problems or to seek help begins with communicating with others.

Communication skills are probably the most important skills we have. We communicate with each other all the time, so most of us have plenty of practice. At work or college, you will be communicating with a wide range of people and in the process becoming part of a working and learning team. It is important to know each person's role and responsibilities within the team so that communication is made as easy as possible.

When you first start work or college, listening to and following instructions will be the most important part of working with others and effectively communicating with them. You will also be expected to keep your supervisor or tutor informed about your work and may have to deal with a number of questions relating to it.

The Disadvantages of Oral Communication

It is probably not what you joined the construction industry for, but it is good to get on with other people at work! We probably spend more time talking and listening to people than using any other method of communication. If it is done effectively, it is usually the most successful method to use.

It gives you direct, personal, contact. You can raise questions, answer them, discuss and settle misunderstandings as they arise.

However, there are disadvantages to spoken communication, some of which are:

- there is no written record, so it could lead to dispute over what was said;
- there is not always enough time to think clearly;
- it is harder to voice criticism;
- it is hard to carry out with large numbers of people.

Communicating Effectively with Colleagues

What does 'communicating effectively with colleagues' mean? Let us first look at some of the terms used.

'Personal communication' means talking to other people. That means speaking, listening and discussing things that come up. 'Colleagues' are the people you work with – your workmates and site managers. If you want to enjoy a productive working relationship with them, you will have to work as a member of a team to get things done – and that means trying to communicate in a positive and clear way with your colleagues/friends/supervisor.

Body Language

We use body language to communicate messages, although in most cases we are not aware of it. And while we are conveying those messages, we can give off unconscious messages about how we're feeling, sometimes about the person we're trying to communicate with. Figure 4.3 shows two very different types of body language. In the first picture, the listener is attentive and open. In the other picture, he gives off the unmistakable message of not wanting to be there. This is why it is important to use the correct body language while we are at work or college.

Tip

Sometimes, communication is difficult if you don't know what certain words mean. Try to look them up in a dictionary or in the glossary of your textbooks. Or you could always ask your supervisor/tutor to explain them to you.

(a)

(b)

Figure 4.3 (a) Positive and (b) negative body language.

We often use hand gestures as well as words to get across what we are saying, to emphasise a point or give a direction. Body language is quick and effective. A wave from a distance can pass on a greeting without being close, and using hand signals to direct a lorry or a load from a crane is instant and does not require any other equipment, such as radios.

Try this Out

Discuss with a colleague what hand signals or gestures you would use on site to give the following urgent messages to people who can see you but can't hear you.

- To tell a colleague that he has an urgent phone call.
- To warn a colleague walking under a roof that rubble is about to be thrown down.
- To tell a lorry driver to stop and reverse immediately
- To warn a colleague to get out of the way of an approaching crane arm.

Tip Try your ideas out on each other to see if the meaning is clear.

Conflicts

No matter where you work, there will always be problems: not enough bricks, mix too wet, someone off sick and so on. Your tutor or supervisor is not a mind reader – you need to tell them if there is a problem or conflict.

Confirming Instructions

While at work or college, you will be told what to do by your supervisor or tutor. You need to listen to their instructions carefully because they may give you several tasks at once. You will need to decide which job to do first.

Information Sources

Always check that information sources such as working drawings or specifications comply with good practice. Your tutor or supervisor will inform you about this and will have a checklist for you to follow.

Discrepancies in Information

If you notice a discrepancy in information such as a working drawing at college or work, make a written note of it so you don't forget it. Then later on you can inform your tutor or supervisor and then they can decide what to do. If you discover a discrepancy such as the wrong measurement on a working drawing, always report it to your tutor or supervisor so they can deal with the matter in a satisfactory way.

Face-to-face Communication

You will need to listen to spoken information and instructions daily in the workplace or at college. This requires good listening skills and practice in remembering details.

Listening Skills

Some information you hear will be more important than other information. You need to:

- look at the person speaking and listen carefully;
- identify your purpose for listening;
- select the important information;
- check and remember the important information.

Speaking Skills

Working in the construction industry requires you to speak to lots of different people, from colleagues, supervisors and tutors to members of the public.

> **Tip**
>
> When complaining about a problem you need to be:
> - polite
> - clear
> - give all the facts
> - give the details.
>
> Use the right body language. The aggressive body language shown in Figure 4.3b just upsets people and will not get things done any quicker.

> **Remember** Following instructions correctly is very important. If you don't understand them, you must ask for more information.

Three main things may affect the way you speak to people at work:

- your relationship with them
- your reason for speaking to them
- communication barriers.

Relationships

You will speak to lots of different people at work and at college. Your relationship with each one of them will be different.

Think of the ways you talk to these people such as your supervisor, tutor, colleague or members of the public. Do you change your language at all? How and why?

Reasons for Speaking

There are various reasons for speaking:

- to pass on information
- to ask for information
- to reply to questions.

Think of examples of each from your college or workplace.

Communication Barriers

There are many barriers to communication:

- surrounding noise
- cultural differences
- language barriers.

How do these affect what we say?
To check that the other person understands you:

- ask questions
- repeat information
- notice the listener's body language.

Try this Out

Think of examples of body language that mean 'I don't understand'. Write them down and then show them to your tutor for assessment.

Programmes of Work

Scheduling operations is an important part of a building project, requiring you to understand the sequence of operations and how this is logged on a chart or project planner. You may be familiar with a range of formats for programmes of work and the need to understand the importance of work not falling behind schedule. This section looks at how to read and understand a typical programme of work and how to amend it following a change.

Keeping to the Schedule

Certain building jobs have to be completed before others can begin. A chart is made that lists all of the jobs that need to be done. It shows the expected start and finish dates for each of the jobs. This kind of chart is called a programme of work. The programme of work tells you when you are needed on site and when different materials and equipment are needed.

The chart can also be used to track progress. There are some empty rows for you to fill in. You record the actual start and finish dates for each of the jobs.

Figure 4.4 shows you a section from a programme of work.

Notes on the use of the chart:

- The chart is based on a five-day working week.
- A week commencing date is shown at the top of each column. It tells you the date of the first day of the week. You use it to work out the dates of the other days in that week.
- Jobs are listed on separate rows.
- You record the actual start and finish dates of the jobs in these rows.
- Blocks of colour are used to show when jobs are planned to start and finish. The number of coloured blocks tells you the number of days the job is planned for (e.g. brickwork to damp proof course is planned to take three days).

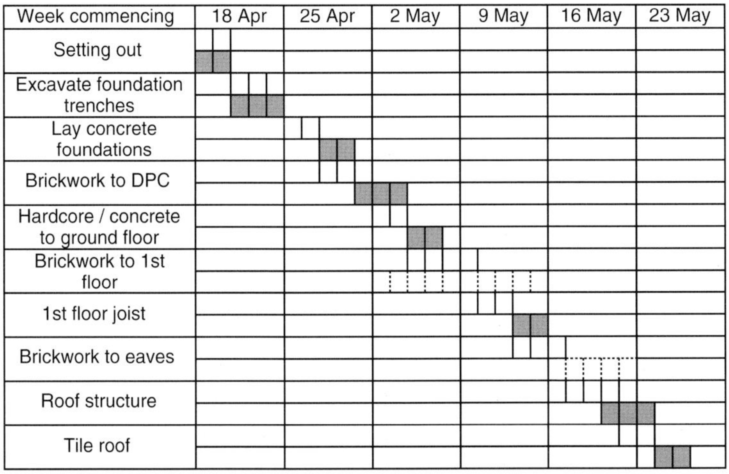

Week commencing	18 Apr	25 Apr	2 May	9 May	16 May	23 May
Setting out	▓					
Excavate foundation trenches	▓▓					
Lay concrete foundations		▓▓				
Brickwork to DPC		▓▓				
Hardcore / concrete to ground floor			▓			
Brickwork to 1st floor						
1st floor joist				▓		
Brickwork to eaves						
Roof structure					▓▓	
Tile roof						▓▓

Figure 4.4 Programme of work.

Planning

Planning and scheduling work is critical in construction, particularly for bricklayers. This section looks at the rounding and estimating skills required to schedule the work.

Although answers are not expected to be exact, rounding and estimating require an understanding of number skills and accurate calculation skills.

Try this Out

Scheduling Your Work

The first lift of scaffolding is in place. You have calculated that the wall area between the first and second lift is 41 square metres. How long will it take to lay 41 square metres of blockwork and brickwork to the second lift if you work alone?

Approximately .... days

An experienced bricklayer can be expected to:

- Lay approximately 30 blocks or three square metres of blockwork per hour.
- Lay approximately 60 bricks or one square metre of brickwork per hour.

How long will it take to lay 36 square metres of blockwork and brickwork to the second lift if you work with one other experienced bricklayer?

Approximately ....day

Words and Meanings

Approximately – Means 'about' or 'roughly'.

Note

A half-day's work is approximately four hours. A whole day's work is approximately eight hours. A week's work is approximately 40 hours.

Try this Out

Despite every effort to keep to a programme of work, jobs do not always run to plan. On the project illustrated in Figure 4.4, rain caused a delay in laying the concrete, causing all the other jobs to start and finish late.

Complete the chart illustrated in to show:

1. That brickwork to the first-floor started two days late and lasted for four days.
2. That brickwork to the eaves started two days late and lasted for three days.

Tip
Each week on the chart represents five working days.

Workplace Communication

A construction site can only work efficiently when all administration and communication documentation are in operation. A clear understanding of the various documents in common use and their effective completion is vital for you to become an efficient team member.

Memos

Memos are used within a company, between individuals and departments.

Memos are usually short and to the point, but they still need to be effective. Remember the reader of the memo will usually be familiar with the topic area, which means you are able to be direct and concise.

Message Requirements

The majority of communication within and between construction companies takes place by means of messages, for example emails. The issuer of the message is able to ask precisely the information required and in the desired order.

When you record a message, remember the following basic rules:

- Complete all dates, times, content and contact name and details accurately.
- Be brief but use formal English.
- Deal with one topic at a time and then in order.
- May be hand- or typewritten.

Fax Machines

Fax (facsimile) machines transmit information from one location to another by turning it into digital impulses and passing them to the receiving machine at the other end of the telephone line. They can be used to transmit letters or other written information very quickly, to locations all over the world. All faxes are capable of both transmitting and receiving documents.

Telephone

The telephone as a means of electronic communication (as are things like emails and faxes) will provide immediate and constant contact with head office, the architect, builders' suppliers and all statutory authorities involved in the building process. An internal telephone system allows managers and supervisors on a large site to coordinate their resources without the delays caused by circulating written information.

Points to remember when using a telephone:

Try this Out

At lunchtime, you take a phone call from the builders' merchants to say that they will not be able to deliver the bricks and blocks that were ordered till tomorrow. Write a memo to your boss, Darren Whatmough, to let him know about the situation so that he can deal with it when he gets back from lunch.

Include in the memo the date, time, content and contact name and details of the call.

The majority of firms have a standard printed memo form, similar to that shown in Figure 4.5.

Academy Construction Co.
Training Road
Esford
PL4 7GH

Memo

To: Site manager   From: _____

Date: _____   Subject: Material change

Figure 4.5 Example of a memo.

- Speak clearly and give the receiver of your call time to take notes.
- Listen carefully.
- Do not shout.
- Give the receiver of your call time to answer your questions.
- Remember: facial expressions cannot be seen.

Try this Out

Telephone Calls

You are a bricklayer working for Joseph Ellam Construction and need to contact Paul Howells about attending a site meeting next Monday at 10 a.m. Unfortunately, he was not in his office when you rang. He has, however, left a message on his answer phone.

Answer Phone Message

'Hello, this is Paul Howells. I'm sorry I can't take your call right now. Please leave a message after the tone.'

Using the information Paul has left you, practise leaving a message on his answer phone.

Make sure you include all the relevant details in your message.

Tips

- Use your own name.
- Use the company name and telephone number.
- Take turns with a colleague to observe each other.
- Record how you did using a checklist.
- Discuss the areas you both think you did well in and those you think could be improved.

Before dialling, make short notes about the key points of the topic; this will help you to remember important information. Whenever possible, find out who you need to speak to before dialling.

Email

This is a communication method that uses the Internet. It enables you and your company to communicate effectively and simply on a worldwide basis. Messages are stored on a network until the recipient accesses them.

Longhand

Longhand is a term used to describe text written by hand (handwriting) as opposed to text that has been typed with a keyboard. You could use longhand to carry out basic calculations and the noting of measurements from working drawings.

Business Letter

A business letter is a more effective form of communication than, say, talking to someone.

A letter:

- is a permanent record that you can store and refer to at a later stage;

- enables you to prepare ideas and organise them in a coherent way;
- effectively conveys detailed information and complex ideas.

A business letter should give a positive impression of you and your company. Any letter should be clear, concise, logically written to enable any reader to understand it and, above all, without mistakes.

Most business letters tend to follow a standard format regarding layout and will be on a company's headed notepaper. You will probably write business letters for the following reasons:

- obtaining information or quotations from other companies;
- placing orders for materials or services;
- making arrangements with other companies;
- seeking advice from consultants and the like.

Business letters need a structure; otherwise, they will not make sense. They require a clear beginning and end, and a logical format for everything in between.

Radio

On large sites, direct communications can be provided by the two-way radio telephone, which enables an operative to move freely around the site and yet remain in contact with the main site office, as well as with other operatives or vehicles similarly equipped.

Complete site coverage can be obtained by distributing handsets to all key personnel, with the control point situated at the main site office.

Note

- Be clear – give all the relevant points and details, so plan it out carefully beforehand.
- Be concise – keep to the point, so do not include unnecessary information.
- Be courteous – write in a polite way and the person you are writing to will look favourably on your request.
- Follow the layout of the business letter shown in Figure 4.6.

Organisational Documentation

Research has shown that the completion and storage of documents and information can save considerable time, effort and paperwork. One way to achieve this is to have site documentation that is designed for simple completion and contains the maximum amount of data necessary to inform relevant personnel about, or to store or update information connected to, a particular project or specification detail.

Examples of company organisational documents can include the following.

Timesheets

A timesheet (Figure 4.7) is a form completed by each member of staff on a weekly basis. Included on the timesheet would be details of hours worked and a description of work carried out.

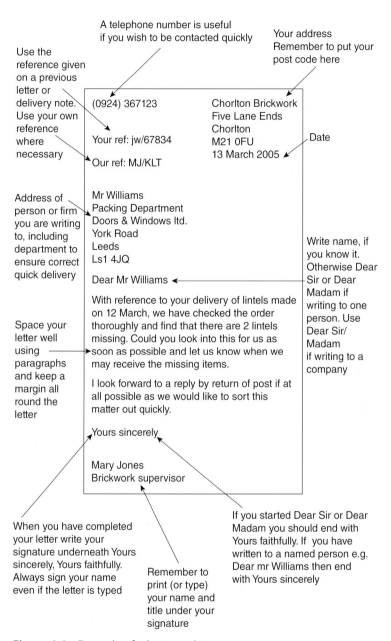

A telephone number is useful if you wish to be contacted quickly

Your address Remember to put your post code here

Use the reference given on a previous letter or delivery note. Use your own reference where necessary

Address of person or firm you are writing to, including department to ensure correct quick delivery

Space your letter well using paragraphs and keep a margin all round the letter

Write name, if you know it. Otherwise Dear Sir or Dear Madam if writing to one person. Use Dear Sir/ Madam if writing to a company

(0924) 367123

Your ref: jw/67834

Our ref: MJ/KLT

Chorlton Brickwork
Five Lane Ends
Chorlton
M21 0FU
13 March 2005

Date

Mr Williams
Packing Department
Doors & Windows ltd.
York Road
Leeds
Ls1 4JQ

Dear Mr Williams

With reference to your delivery of lintels made on 12 March, we have checked the order thoroughly and find that there are 2 lintels missing. Could you look into this for us as soon as possible and let us know when we may receive the missing items.

I look forward to a reply by return of post if at all possible as we would like to sort this matter out quickly.

Yours sincerely

Mary Jones
Brickwork supervisor

When you have completed your letter write your signature underneath Yours sincerely, Yours faithfully. Always sign your name even if the letter is typed

Remember to print (or type) your name and title under your signature

If you started Dear Sir or Dear Madam you should end with Yours faithfully. If you have written to a named person e.g. Dear mr Williams then end with Yours sincerely

Figure 4.6 Example of a business letter.

Name:		NVQ No:	
Contract:		W/C:	

Day	Activity	Hours	Difficulties
Mon			
Tue			
Wed			
Thu			
Fri			
Sat			
Sun			

Figure 4.7   Example of a timesheet.

Day Worksheets

Day worksheets (Figure 4.8) are sometimes confused with timesheets in that they are thought to carry out the same purpose. This is not the case. Day work is carried out without an estimate. Day worksheets record work done, hours worked and usually materials used.

Job Sheets

A job sheet (Figure 4.9) or card is a simple form allocating a job or jobs to a person for a specific day or week.

Variation Order

Used by an architect to make any changes to the original working drawing, a variation order (Figure 4.10) includes omissions, alterations and extra work.

Delivery Records

Delivery records (Figure 4.11) list all deliveries over a certain period and are sent to the builder's main office so that payment can be made.

Delivery Notes

Delivery notes are given to the builder by the supplier, and list all the materials and components being delivered.

Confirmation Notice

A confirmation notice (Figure 4.12) is given to the builder usually by the architect to confirm any changes made in the variation order.

Sheet no. ....

Job title ....

Week commencing ....

Registered office

Description of work					
Labour	Name	Craft	Hours	Gross rate	Total
			Total labour		
Materials		Quantity	Rate	% Addition	
			Total materials		
Piant		Hours	Rate	% Addition	
			Total plant		
Note	Gross labour rates include a percentage for overheads and profit as set out in the contract conditions		Sub total		
			VAT (where applicable)___%		
			Total claim		
Site manager/foreman					
Architect					

Figure 4.8 *Example of a day worksheet.*

Invoices

State what has been provided from various sources and how much the builder will be charged for it.

Order Forms

Order forms (Figure 4.13) are used to order materials or components from a supplier.

Job Sheet				
Name:_____ Week commencing:_____				
Site:_____ Foreman:_____				
Description	Expected time	Resources	Difficulties	Remedial action required

Figure 4.9   Example of a job sheet.

Variation to proposed works

Reference no:

Date _____

From _____

To _____

Possible variations to work

Additions

Omissions

Signed _____

Figure 4.10   Example of a variation order.

Customer ref _____

Customer order date _____

Delivery date _____

Item no	Qty supplied	Qty to follow	Description	Unit price

Delivered to:

Customer signature: ....

Figure 4.11 Example of a delivery record.

Site Forms

Most bricklayers do not complete many forms on a day-to-day basis. Most of the forms you encounter will relate to your employment status, health and safety issues or security. These are important forms and need to be completed accurately (Figure 4.14).

On all sites, it is important that the site manager knows who is on site and who to call in case of emergency. On large sites you may need a site pass to get on and off site. You need to be accurate when filling in forms.

Filling in forms may not be exciting, but it is necessary. If your personal details are going to be put on an ID card, you want them to be accurate.

Confirmation for variation to proposed works

Reference no:

Date _____

From _____

To _____

I confirm that I have received written instructions
From   _____
Position _____
To carry out the following possible variations to the above named contract

Additions
Omissions

Signed ....

Figure 4.12   Example of a confirmation notice.

Daily Report or Site Diary

A weekly or daily report (Figure 4.14) is used to convey information such as weather reports, attendance and so on to head office and to provide a source for future reference.

Accident or Near-Miss Reports

All accidents and near misses must be reported by site personnel to their supervisor and recorded (see Chapter 1). It is a legal requirement for details of an accident or near miss to be recorded in the Accident Book, which must be kept on each construction site for that purpose.

Materials order form			
Order number		Date	
Site address			
Name/address of supplier			

Please supply the following order to the above address:

Description	Quantity	Date required

Special delivery instructions:

Signature of site manager	

Figure 4.13 Example of an order form.

bardsley

Bardsley Construction Limited

WEEKLY PROGRESS REPORT	Contract: **Rochdale 6th Form College**	Week ...38 of 70......wks w/e 8/1/2010..............
Start Date 24/04/09 Planned Completion Date:20/08/10		Programme minus 4 week (Last Report) minus 2 week
Labour on Site	Weather:0 hours	Frost:/snow40 hours
	Rain to date:301 hours	Frost to date:70 hours

	S	S	M	T	W	T	F		S	S	M	T	W	T	F
Labourers			4	3	2	2	3	SPS TANKING							
Welfare g/f	1			1	1	1	1	Roof Cladders			3				
Ground-Workers			1	1				Joiners			3			3	3
Bricklayers			6	6				Sidlow concrete							
Hod Carriers			2	2				Electricians/turners						1	
Scaffolders			4					Plumbers/ameon			3	5	5		
Steel erectors								Supyk drylining			114	1			
Ventilation ductwork								painters							
PC stairs/picles								mechanical			3	4	4		
Floor decks								accufix			5				
F/LT + Opert			1	1	1	1	1	PRG windows			8				

Subcontractors:

DCT> site overstone tidy
Supyk drylining
 Falcon roof decking
PRG windows & curtain wall
IBN scaffold h/l floor handrails & roof
Joiner > shuttering to deck levels
Bricklayers
Ameon plumbing & mechanical

Materials

    All as buyers calloff orders
    Meeting Tuesday with internal door supplier

Service Authorities:

Figure 4.14 Site forms.

bardsley

Bardsley Construction Limited

Build control local authority attended site all good order
Aegis COW now visiting site weekly (Peter Smith)

Labour Requirements

None reqd. at present

Health & Safety Issues

Site safety audit actioned as of 2/12/09 byBCL Dan Taylor achieving inspection score of 93%
All remedial actions undertaken by w/c 7/12/09

Simian Risk now actioningsite visits for weekly scaffold inspections

Any Other Matters:

Due to adverse weather condition site non operable for w/c 4/1/2010 all week
also bad weather prevented recovery works to programme for xmas week
 In all site suffered severe weather conditions from 15/12/09

Figure 4.14 Continued

Summary

A building site can only work efficiently when all administration factors are in operation. If one of those factors is missing, it may well affect the efficiency of the entire building process.

Information

For more information about communication and site documentation, visit the websites of the better-known building companies.

Quick Quiz

1. What information would you obtain from a programme of work?
2. What is the difference between a day worksheet and a timesheet?
3. What information is required on a job sheet?
4. Describe three methods of communication.
5. List three types of electronic communication.

CHAPTER 5

Mixing Mortar

THIS CHAPTER RELATES TO UNIT CC1003K AND UNIT CC1003S.

Words and Meanings

Aggregate – The inert material forming the body of the mortar, sand being normally used for this purpose.
Matrix – The material which, when combined with water, hardens and binds the aggregate together. The matrix may be either lime or cement, or a mixture of the two.

Mortar is a material which is used to bond bricks and blocks together to form a solid unit of masonry. Mortar is composed of a matrix and an aggregate which, when mixed with a certain quantity of water, become sufficiently plastic to be spread in thin layers. In a short time, varying with the matrix used, the mortar sets and gradually hardens.

The Components of Mortar

Sand

Sand for bedding brickwork is used as an aggregate in preparing mortar. Sand used in mortar for brickwork should be free from all earthy and organic matter and not too fine. While it should generally be fairly sharp or gritty, an addition of soft sand will make the mortar easier to manipulate.

Silt Test

Sand can be tested for cleanliness by carrying out a silt test (Figure 5.1) on a small quantity of sand taken from a load. The amount of silt and clay should not be more than 10% of the volume of aggregate.

Cement

Cement for bedding brickwork is used largely as the matrix in preparing mortar. Cement is manufactured from chalk (calcium carbonate), which is excavated from deposits of chalk, then heated in a long, rotating kiln. The clinker derived from this process is then ground to a fine powder and bagged for distribution, usually in 25 kg bags (Figure 5.2).

A simple experiment to test for the cleanliness of sand

The silt test:

Sand can be tested for cleanliness by carrying out a silt test on a small quantity of sand taken from a load.

Material and equipment required:

- Glass measuring cylinder or jar
- Sample of sand
- Ruler or tape measure
- Water

Method:

Half fill the jar with water and add one teaspoon of salt.

Add 100 mm of sand.

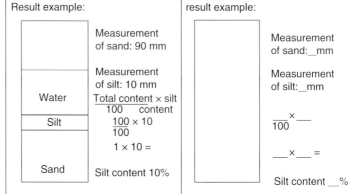

Result example:

Water

Silt

Sand

Measurement of sand: 90 mm

Measurement of silt: 10 mm

$$\frac{\text{Total content} \times \text{silt}}{100} \text{ content}$$

$$\frac{100 \times 10}{100}$$

$$1 \times 10 =$$

Silt content 10%

result example:

Measurement of sand:__mm

Measurement of silt:__mm

$$\frac{\_\_}{100} \times \_\_$$

$$\_\_ \times \_\_ =$$

Silt content __%

Silt test diagram

The cylinder is then shaken vigorously, and the contents allowed to settle for three hours. Sand containing more than 8% of silt should not be used for mortar as it will greatly reduce the bond between the brick and the mortar. The above diagram shows an example of a completed silt test.

Results of experiment:

The sample of sand should not contain more than 8% silt. The sample of sand tested was suitable/was not suitable for use in concrete mix. (Delete as appropriate)

Figure 5.1 Silt test.

Figure 5.2 Cement.

Lime

Lime for bedding brickwork is used largely as a matrix in preparing mortar. Lime is produced by the burning of calcium carbonate in the form of chalk, limestone or marble, the result being calcium oxide or quicklime; for building purposes, quicklime requires slaking.

Manufacturers now supply lime as a hydrated lime in the form of a dry powder. This hydrated lime is ordinary quicklime that has been slaked by the combination of just sufficient water to break it down from lumps to a dry powder. If properly stored, it remains dry and is ready for immediate use by mixing with water (Figure 5.3).

Water

All water used in the mixing of concrete and mortar should be clean and free from organic or mineral impurities.

Additives

There are a number of proprietary additives that are used when mixing concrete or mortar for a variety of purposes (Figure 5.4). One such is plasticiser, which when added to the mixing water makes the concrete or mortar more workable and frost-resistant.

Figure 5.3   Lime.

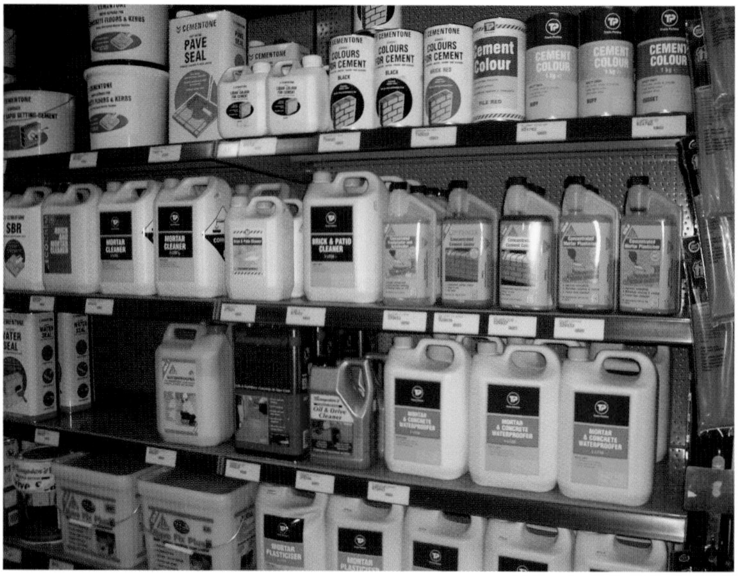

Figure 5.4 Additives.

> **Note**
>
> Cement is an irritant and care must be taken when handling it to avoid skin contact, breathing in the dust and cement particles or getting it into the eyes. Personal protective equipment (PPE) is essential.

Other additives include accelerators, to make mixes set more quickly; retardants, to slow down the setting process thus enabling more time to work the concrete and give a harder set; and pigments, to give a range of colours to mortars and concretes.

Mortar Mixes

Mortar is prepared by the following methods.

Lime Mortar

This usually consists of one part lime to three parts sand, but may vary in its proportions according to the type of work for which it is intended.

If slaked quicklime is used, the method of preparation will differ from that required for powdered hydrated lime. In both cases a platform on which the materials may be mixed is required, the purpose being to provide a flat level surface and to keep the mortar free from foreign matter that would be unavoidably collected if mixed on the ground.

Mixing by Hand

When hydrated lime is used, being in powdered form, it is more convenient to mix the lime with the sand before any water is added. The sand is placed in a heap to one side of the platform and over this is spread the lime (Figure 5.5), the two materials being then repeatedly turned over with shovels until a thorough mix is ensured (Figure 5.6).

Figure 5.5 Sand and lime heap.

Figure 5.6 Sand and lime heap: mixing with a shovel.

With the two materials thus mixed, a trough or hole (usually called a ring) is formed in the middle of the heap of lime and sand (Figure 5.7), water is added and the following procedure performed: the sand is gradually worked from the inside of the ring into the slurry (Figure 5.8), every endeavour being made to maintain the ring until mixing is almost complete.

Water is added during the process as required to reduce the mixture to a fairly pliable state (Figure 5.9).

Since there is always a possibility of small particles of lime being unslaked, it is the general practice to allow lime mortar to stand for at least 24 hours before use.

Cement Mortar

This is a mixture of Portland cement and sand in proportions that will vary with the nature of the brickwork to be built. For all general purposes this is one part cement to four parts sand.

The mixing of cement mortar follows the method outlined for hydrated lime mortar. With cement as a matrix, the mortar begins to set about 30 minutes after the water has been added, and because of this it should be prepared in fairly small quantities and disturbed as little as possible after setting has commenced.

Figure 5.7   Sand and lime heap: ring and water.

Figure 5.8   Sand and lime heap: turning over.

Figure 5.9   Sand and lime heap: mixture to pliable state.

Gauged Mortar

It is now quite common to use a combination of lime and cement as a matrix with the usual aggregate of sand, this being known as gauged mortar. There are two methods of introducing the cement to the mixture:

Method One

Lime mortar is prepared in bulk, as described earlier, and from this sufficient is set aside for immediate use. To this lime mortar, cement is added as required, the usual proportion for general work being ten per cent of the whole.

Method Two

In this case all dry materials are mixed as required with one part cement, two parts hydrated lime and six parts sand. The operation of mixing is similar to that described for hydrated lime or cement mortar. It must be remembered that since cement has been added it should only be prepared in small quantities.

Apart from the fact that lime/cement mortar is much easier to manipulate than a cement mortar, it also has the advantage of rendering the brickwork more damp-proof. Investigations by the British Standards Institute proved that cement mortar joints tend to shrink away from the surface of the brick, leaving capillary channels through which water may find access, while a gauged mortar will form a much better seal.

It appears therefore that, unless great compressive strength is required, gauged mortar is to be preferred for all general building purposes. For small quantities of mortar, mixing by hand as required is the more suitable method of preparation, but for large quantities machine mixing is generally adopted.

Mixing by Machine

There are two common types of mortar mixers:

1. the revolving drum
2. the mortar pan.

Revolving Drum Mixer

This consists of a drum (Figure 5.10) with an open end and is obtainable in various sizes according to the number of bricklayers to be supplied with mortar. Fitted to the inside of the drum are blades that mix the materials as the drum revolves. The matrix and the aggregates are placed in the drum with sufficient water to make pliable mortar when the drum is set in motion. The drum is revolved for a period of time, after which the mortar is ready to use.

Mortar Pan

This is a large pan pivoted on a central axle, with heavy rollers mounted inside the pan (Figure 5.11). With the pan revolving in an opposite direction to the rollers, the matrix and the aggregate are added, together with the necessary amount of water. The action serves to knead and mix the matrix and the aggregate, and to break down any large particles in the latter.

It should be noted that:

● Cement mortar is generally used where maximum strength and durability is required and where settlement is to be avoided. As

Figure 5.10 A revolving drum mixer.

Figure 5.11 A roller pan mixer. Courtesy of Colin Freeman, Terjon Plant and Engineering Ltd.

it is quick-setting, it allows rapid progress in the building of brickwork.

- Gauged mortar is an excellent mortar for general use, with very little loss in strength as compared with cement mortar.
- Lime mortar is much weaker than cement mortar but is used principally because it is smoother and easier to handle, and less expensive, while producing sound and dependable brickwork.

Pre-mixed Mortars

These are mortars which are produced in a quality-controlled environment to ensure accurate mix proportions. There are two main types available:

Ready-to-use-mortars

These mortars are produced in a factory and delivered to site ready-to-use. They may be:

- wet ready-to-use, which requires no further mixing and is stored in tubs on site (Figure 5.12);
- dry ready-to-use delivered in silos or bags, which requires only the addition of mixing water.

Figure 5.12 Example of a mortar tub.

Lime/Sand Mortars

These are pre-batched materials which are delivered to site, with cement and water being added prior to use.

All production methods of factory-produced mortars can offer both coloured and natural shades.

Mortar Silos

Factory-produced silo mortars offer a range of mix proportions and should overcome any potential problems related to on-site mixing.

Method

A silo (Figure 5.13) is delivered to the site complete with internal mixer. Once power and water supplies are connected, mortar can be produced as required. Water is added to produce the necessary consistency.

Two-compartment Silo

This is a movable silo with two sealed compartments that are filled with the required amounts of sand and cement. The mixing ratio is decided on before delivery to ensure that the mix proportions are to the customer's requirements.

Figure 5.13 Example of a silo.

Single-compartment Silo

These are single-compartment movable silos that are filled by the mortar manufacturer with dried sand, cement or lime, if required, and other admixtures, pigments or additives that are premixed to the customer's requirements.

Water and pigment are added to both types of silo by means of a metered pump.

Information

For more information about mortar, cement and limes, visit the websites of the better-known manufacturers and suppliers, or the Mortar Industry Association's website: www.mortar.org.uk.

Quick Quiz

1. Describe two types of mortar silo.

2. What does a silt test measure?

3. When mixing by machine, what should be added to the mixer first?

4. From what material is cement produced?

5. What is a retardant used for?

CHAPTER 6

Roof Construction

THIS CHAPTER RELATES TO UNIT CC1003K AND UNIT CC1003S.

Roofing Construction: Types of Roofs

Although there are several different types of roofing, all roofs will technically be either a flat roof or a pitched roof.

Flat

A flat roof is any roof which has its upper surface inclined at an angle (also known as the fall, slope or pitch) not exceeding 10 degrees (Figure 6.1).

A flat roof has a fall to allow rainwater to run off, preventing puddles forming as they can put extra weight on the roof and cause leaks.

Pitched

There are several different types of pitched roof but most are constructed in one of two ways: prefabricated or traditional.

Prefabricated

A trussed roof is a prefabricated pitched roof specially manufactured prior to delivery on site, saving timber as well as making the process easier and quicker. Trussed roofs can span greater distances without the need for support from intermediate walls (Figure 6.2).

Traditional

Traditional roofs are entirely constructed on site from loose timber sections using simple jointing methods (Figure 6.3). This type of roof uses rafters that are individually cut and fixed in place, with two rafters forming a truss. Once the rafters are all fixed, the roof is finished with felt, battens and tiles.

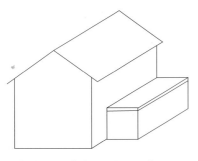

Pitched roofs (more than 10°)
flat roof (10° or less)

Figure 6.1   Pitched and flat roof.

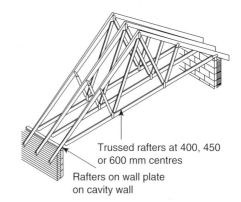

Trussed rafters at 400, 450
or 600 mm centres

Rafters on wall plate
on cavity wall

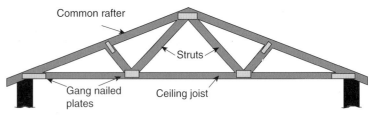

Common rafter

Struts

Gang nailed
plates

Ceiling joist

Figure 6.2   Trussed roof construction and trussed roof rafter.

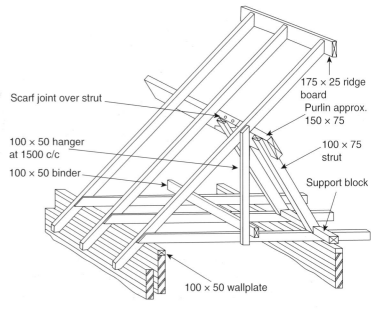

Scarf joint over strut

175 × 25 ridge
board
Purlin approx.
150 × 75

100 × 50 hanger
at 1500 c/c

100 × 75
strut

100 × 50 binder

Support block

100 × 50 wallplate

Figure 6.3   Traditional roof construction.

Loadings

When walls are carrying floors, deflection in the floor will concentrate the load on the inner edge of the wall. This eccentric loading, as it is termed, should be avoided as much as possible because it has a marked effect upon the strength of the walling.

An example of eccentric loading is the transmission of loads from the roof to the external walling. The loads are usually distributed along the length of the wall by means of a wallplate placed on the inner leaf of a cavity wall, or on the inside face of a solid wall. With masonry walls, this method is satisfactory as the masonry is usually strong enough to resist bending, but when lightweight materials are used for inner leaves of cavity walls, the wallplate should be placed centrally over the wall, as illustrated in Figure 6.3. This will ensure that the load is taken by both leaves and that the wall ties will play little part in strengthening the masonry.

Roof Fixings

Straps

In most instances the roof framework will be required by Building Regulations to give stability to gable walls. This can be best achieved by the correct bracing of the trussed rafters and the provision of adequate connection between the masonry and the trussed rafters.

This connection takes the form of 2 m long galvanised steel straps fixed over at least three trusses and then built into the inner leaf of the cavity wall and turned through 90 degrees into the cavity.

They should be fixed on the underside of rafters and on the top side of the ceiling joist. The distance apart should not exceed 2 m and they should be fixed to the timber member with a minimum of four screws to each timber member.

Wallplate Straps

Wallplate straps are components, usually made from pre-galvanised strip mild steel or stainless steel, which provide vertical restraint and secure the wallplate to the inner leaf of a masonry wall (Figure 6.4).

Figure 6.4 Restraint straps.

Roof Components

Figure 6.5 illustrates many of the components for roofing. These include:

- **Ridge:** The ridge acts as a spine at the apex of the roof structure, running horizontally and against which the uppermost ends of the rafters are fixed.

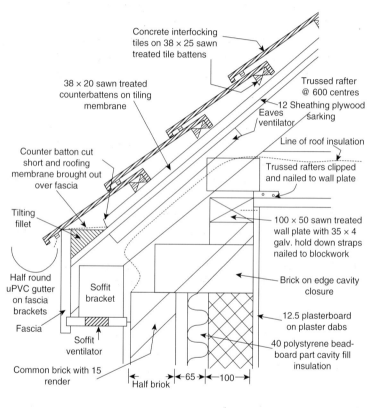

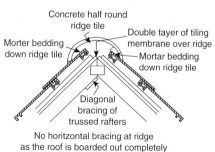

Figure 6.5   Roof section details.

- **Purlin:** Purlins are horizontal beams that support the rafters midway between the ridge and the wallplate when the rafters are longer than 2.5m.
- **Firrings:** A firring is a long wedge, tapered where joists are parallel to the fall, or of variable depth for joists at right angles.
- **Battens:** Battens are narrow strips of timber used in roofing to support and hold the fastened slates or tiles.
- **Tiles:** Tiles are individual roof coverings placed and fastened on battens to provide a weatherproof covering.
- **Fascia:** A fascia board usually measures 175mm × 20mm of 'plane all round' timber is fixed to the vertical cuts of the rafters at the eaves to provide a finish and a fixing board for guttering.
- **Wallplates:** Wallplates are usually made of timber, laid flat and bedded on mortar, running along a wall to carry the feet of all rafters and ceiling joists; they are anchored down with restraint straps to prevent movement.
- **Bracings:** Bracings are strengthening pieces of metal or timber used in the roof's construction.
- **Felt:** Felt is waterproof material available in rolls of various widths and length that is placed underneath battens to provide a water-proof covering.
- **Slate:** Slate is a natural substance worked into specific sizes and weights and placed and fastened onto battens as a waterproof covering.
- **Flashings:** A building technique used to prevent leakage at a roof joint. Normally metal (lead, zinc or copper) but can be cement, felt or a proprietary material.
- **Soffit:** A soffit is the under-surface of eaves, balconies, arches and so on.

Information

For more information on roof construction and components, visit the websites of the better-known roof and component manufacturers.

1. What are the two main ways roofs are constructed?

2. Technically, how many types of roof are there?

3. Explain the term 'eccentric loading'.

4. List five roof components.

5. What is the purpose of wallplate straps?

CHAPTER 7

Sketching and Drawing

Drawing Equipment

Traditional drawing skills have been used for many years and are an important part of the building process. As bricklayers are expected to work from drawings on site, you must be able to prepare basic outline drawings to scale. To begin with, a good-quality set of drawing equipment is required when drawing to scale (Figure 7.1). It should include:

- **Set squares:** Two set squares are needed: one at a 45 degree angle and the other at angles of 30 and 60 degrees. These are used to draw vertical and horizontal lines.
- **Protractors:** A protractor is used for measuring and setting out angles.
- **Compass and dividers:** Compasses are used to draw circles and arcs of circles. Dividers are used for transferring measurements and dividing lines.
- **Scale rule:** A scale rule with the following scales is required: 1:5/1:50, 1:10/1:100, 1:20/1:200, 1:250/1:2500.
- **Pencils:** Two pencils are needed: HB for printing and sketching and 2H for drawing.
- **Eraser:** A rubber eraser is needed for alterations or corrections to pencil lines.
- **Drawing board:** Drawing boards are made with a smooth flat surface, with edges truly square and parallel.
- **T-square:** A T-square is used for drawing horizontal lines.

You should, whenever possible, produce your own scaled working drawings. In the early stages of learning, graph paper may be used to

Try this Out

Create a scale drawing of the brick workshop, including details such as windows and doors, internal walling and components. Write a report explaining why this technique is still used when computer-aided design is available.

Create a scale drawing of the ground floor of your house. Include such details as walls and doors. A key can be added. Record how long it took to produce the drawing. Would it have been quicker using a computer-aided design system? Explain why you came to the conclusion that you did.

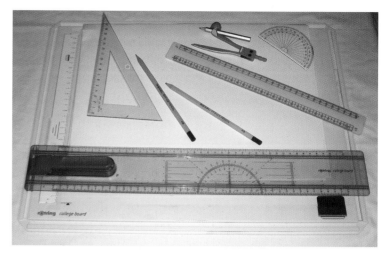

Figure 7.1 Drawing board and equipment.

assist you in correct scaling. It is important that all working drawings are neat and clean.

Using a Scale Rule

Remember Always look at the scale written on plans. You can then use the correct scale rule to read off unmarked measurements accurately.

It can be very helpful to use a scale rule (Figure 7.2) to deal with preparing and reading scale drawings. There are different scale rules that can be used depending on the size of scale you need.

If you have measured something on a plan that reads as 50 mm and the scale used is 1:100, using the 1:100 scale rule you can see this is 5 m or 5000 mm in full size.

1:1	0	10mm	20	30	40	50	60	70	80	90	100	110	120	130	140	150mm
1:100		1m	2m	3m	4m	5m	6m	7m	8m	9m	10m	11m	12m	13m	14m	15m

BS 15cp part 3   315 pl   Magic

1:200	30m	28m	26m	24m	22m	20m	18m	16m	14m	12m	10m	8m	6m	4m	2m	0
1:200	3000mm	2800	2600	2400	2200	2000	1800	1600	1400	1200	1000	800	600	400	200mm	

Figure 7.2 Scale rule.

Drawings and Sketches of Construction Details

The ability to read and understand drawings and sketches is an essential function of any worker in the construction industry. You must study the illustrations in this section carefully to understand the way sketches are drawn. To develop the skill of interpreting these drawings and sketches, you must practise reading sample sketches provided by your tutor.

Drawings and sketches are pictures or diagrams used to describe the construction specifications and procedures required to erect buildings (Figure 7.3).

Construction drawings are generally prepared by architects or engineers using drafting instruments or computer-aided design. These drawings along with written specifications accurately describe the details of buildings.

Sketches are freehand pictures that can be prepared by anyone as a means of explaining specific instructions (Figure 7.4).

In shuttering work, sketches may be prepared by the carpentry foreman to describe the size requirements of a shuttering component or an installation procedure. Look at the sketches and try to identify the shuttering details (Figure 7.5).

Sketches may be prepared using a two-dimensional technique. Figure 7.6 shows the object as it would appear if you were standing

Words and Meanings

Shuttering – Timber or steel casing used to contain concrete while it hardens.

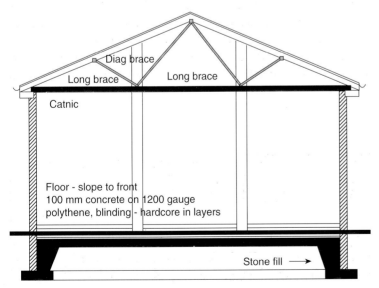

Figure 7.3 Example of construction specifications: a section view.

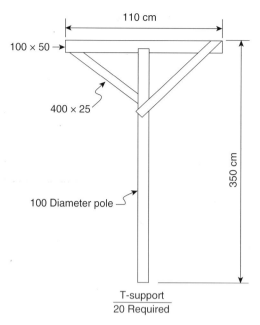

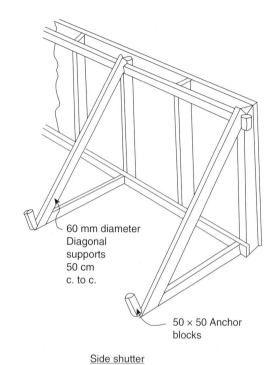

Side shutter

Figure 7.4 Example of a sketch: a T-support.

Figure 7.5 Example of a sketch: a side shutter.

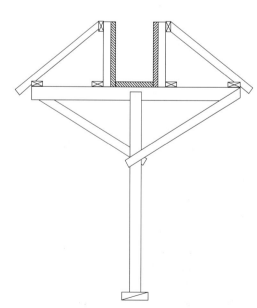

Figure 7.6 Example of a two-dimensional sketch: end elevation of shuttering for a beam.

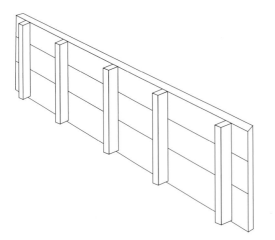

Figure 7.7 Example of a three-dimensional sketch: side shutter panel.

Figure 7.8 Dimension and extension lines.

directly in front of the object. It is a view of the end of a beam shuttering frame. The advantage of this form of sketch is that it is easy to draw. Its disadvantage is that it is more difficult to understand because it does not show the depth, or third dimension, of the object.

Sketches that are prepared using a three-dimensional technique are generally easier for the untrained person to understand. This is because the picture tries to duplicate exactly what the eye will see in real life. Figure 7.7 illustrated shows a three-dimensional sketch of a side shutter panel.

In order that a sketch may be used to instruct a person on the specifications or installation procedures for a shuttering component, it is necessary to attach notes to the sketch.

The size of a component is specified by using a dimension line (Figure 7.8) with an arrow at each end that points to the limits of the measurement.

For clarity, the dimensions are placed around the outside of the sketch. Extension lines (Figure 7.8) are drawn out from the object to indicate the limit of the measurement.

At times, it is necessary to specify the name and/or size of a component part. The use of dimension lines may confuse the picture. In this case a leader line (Figure 7.9) is used to point to the component part and the written specification is given beside the picture. The illustration shows the battens specified beside the sketch and a leader line pointing to the batten in the sketch.

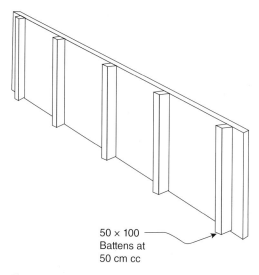

50 × 100
Battens at
50 cm cc

Figure 7.9   Leader lines.

Symbols and Abbreviations

Symbols and abbreviations are used on sketches to reduce the amount of writing required and to make the sketch easier to read. Graphical symbols are small standard pictures used to reduce the amount of drawing detail required on drawings. Abbreviations and graphical symbols are often used together to give complete information. Figure 7.10 shows some of the more common graphical symbols.

Hatchings

Hatching is the term given to the markings on a cross-sectional drawing used to indicate the material that it is constructed of. They are official British Standards and can be found in BS 1192, which controls drawing practice across all sections of the building industry.

Some examples of hatchings are shown in Figure 7.11.

Abbreviations

Abbreviations are a simple way of conveying information on drawings, reducing words to initial letters, for example 'rain water pipe' becomes RWP. They allow the maximum information to be included on the drawing in a concise way.

Abbreviations used in brickwork include:

- airbrick – AB
- brickwork – bwk

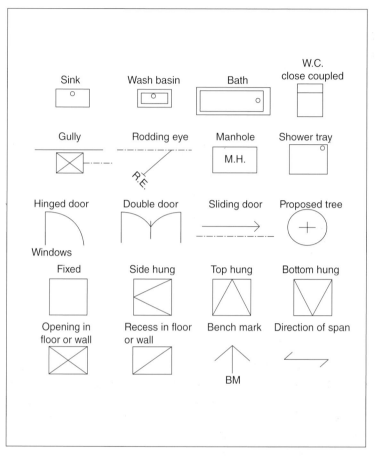

Figure 7.10 Examples of symbols.

- building – bldg
- cement – ct
- column – col
- concrete – conc
- damp proof course – DPC
- damp proof membrane – DPM
- drawing – dwg
- foundation – fnd
- hardcore – hc
- insulation – insul
- reinforced concrete – RC.

Units of measurement are abbreviated as follows:

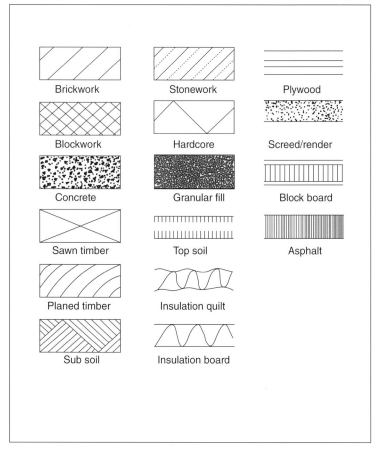

Figure 7.11 Examples of hatchings.

- millimetres – mm
- centimetres – cm
- metres – m.

Sometimes when all dimensions are of the same unit of measurement the unit symbol is left off or a note is placed beside the sketch.

The centre of a hole or bolt is indicated by a centre line symbol as shown in Figure 7.12.

When specifying the distance between parts that are spaced at equal intervals, the symbols c.c. or o.c., meaning 'centre to centre' or 'on centre' respectively, are used.

When specifying the size of a circular part (Figure 7.13) the abbreviation DIA is used to represent diameter. Sometimes the symbol Ø (a circle with a diagonal slash mark through it) is also used.

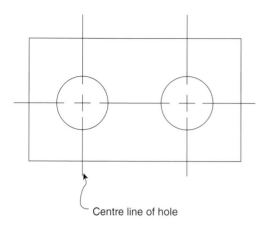

Figure 7.12   Centre line.

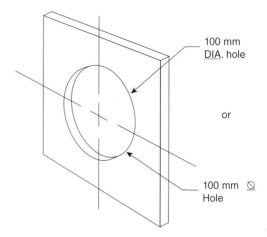

Figure 7.13   Examples of specifying sizes: the circular part.

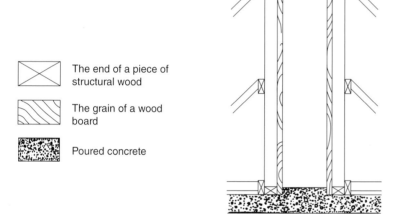

Figure 7.14   Example of symbols: elevation of a section of shuttering.

In order to assist the clarity of a sketch, parts of the sketch may be shaded in to provide contrast. Special symbols are used to indicate the type of material. Figure 7.14 shows an end view of a section of wall shuttering and the material symbols that are used.

To help you find out what different technical words and abbreviations mean, I have listed some tips below.

Using A Dictionary

There are a number of ways you can find out or check the meanings of technical words and abbreviations.

Words and Meanings

Abbreviation – Shortened form of written words, usually the first letter or letters of words, for example on a working drawing DPC may be written in place of 'damp proof course'.

Dictionaries are useful for looking up the meanings of everyday words and abbreviations. For example, 'max' is short for 'maximum'. It means the largest possible size.

Using A Glossary or Website

A specialist glossary is the best place to find out the exact meaning of technical terms, such as masonry units or wall ties.

You may find additional information by looking on a specialist website; a number of these are included in this text.

Remember If you are in doubt about words or abbreviations on a working drawing, ask a colleague or tutor for help. It is very important to get it right. Mistakes cost time and money.

Try this Out

Drawings

Ask your tutor for sample drawings to look at. When you feel confident about reading drawings attempt to draw a range from the list below and show them to your tutor for assessment.

- a cross-section of a concrete floor
- a stretcher bond wall

- a Flemish bond wall
- a cross-section of a load-bearing wall
- a cross-section of an internal blockwork wall
- a cross-section of metal stud partition walling.

You could also have a go at drawing some basic roof components, such as a ridge, purlin, firings and batten.

Information

For more information about drawings and sketches, borrow from your college or local library textbooks that deal with the topic area.

Quick Quiz

1. List five items of drawing equipment and say what they are used for.

2. What is a scale rule used for?

3. What is the difference between a sketch and a drawing?

4. What does the term 'hatching' mean?

5. Why are symbols and abbreviations used on working drawings?

CHAPTER 8

Blocklaying Skills

THIS CHAPTER RELATES TO UNIT CC1014K AND UNIT CC1014S.

Bonding Blockwork

Blocks should be laid out at half bond as far as possible. It will be found useful, when setting out blockwork, to run out the blocks dry first from one end of the wall to the other to ascertain the bond. There will usually be a cut piece of block required to complete the course the same as is found in brickwork.

By beginning the next course from the opposite end of the wall from where you started, with a full block, and working back to the first corner, cutting will be curtailed to a minimum and a satisfactory bond found. This method should, however, not be carried to an extreme, a lap of 150 mm being the very least that should be used.

> **Remember** A regular bond pattern should be maintained throughout the wall's length, ensuring a minimum overlap of a quarter of a block.

Bonding Corners

In order to maintain half bond at junctions and returns, a 100 mm cut block must be built in next to the return block (Figure 8.1). One can choose to bond a blockwork corner in several ways. Figure 8.2 illustrates one method. This example uses a cut block on the second course at the corner. This results in a second cut at the stop end with a smaller cut on the first course.

The illustrations show two methods. There are others; providing that one has kept to the principles of bonding (no straight joints and bonded at the corners), these will be acceptable.

Figure 8.1 shows what is probably the most common method of bonding a blockwork corner. The cut blocks are shaded for emphasis. The most common width for blocks used in construction is 100 mm, the same width as the cut identified. The cut shown in Figure 8.1 acts, as

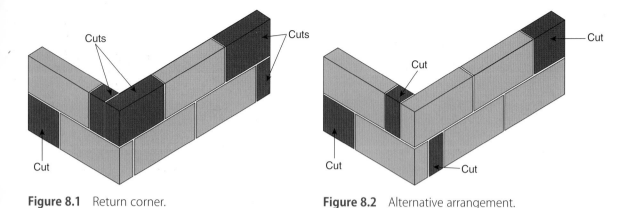

Figure 8.1 Return corner.

Figure 8.2 Alternative arrangement.

it were, as a closer, but if we increase the width of our blocks, what would be the result of this to the bonding arrangement?

The answer, of course, would be that the cut would reduce in length. Figures 8.3 to 8.6 illustrate this point over a range of examples, from partition blocks measuring 50 mm wide to 215 mm load-bearing block-work. The length of the cut reduces as the block width increases. On the 215 mm block corner, the proportions are that of a brick (the width of the block is half of a block's length); therefore, no cut is required.

Bonding Partition Walls

Blocks are ideally suited for the building of both load-bearing and non-load-bearing internal walls. The use of blocks creates a more robust partition, adding rigidity to the whole structure, making it less prone to damage than, say, wood and plasterboard studwork.

External walls that are to receive partition walls must be so built that the eventual tying or bonding in of the partition may be easily carried out. It is therefore necessary that the position of all partition walls should be set out when the main walls have been started, in order that recesses may be formed into which the partitions will eventually be bonded.

Bonding Junctions

As with returns in blockwork, the minimum overlap at a T-junction should be a quarter of a block. This is achieved by building in the 100 mm cut block on alternate courses, as illustrated in Figure 8.7.

The degree of cutting blocks will be determined by the position of the indent in relation to the blocks in the main wall, as illustrated in Figure 8.8. Therefore, bonding junction walls in blocks is similar to bonding junctions in brickwork, and similar issues emerge.

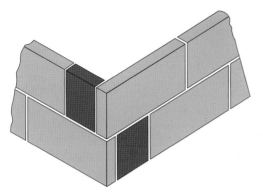

Figure 8.3 50 mm corner.

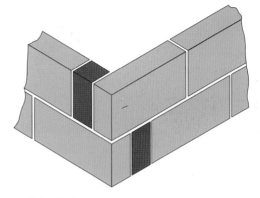

Figure 8.4 100 mm corner.

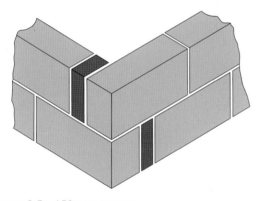

Figure 8.5 150 mm corner.

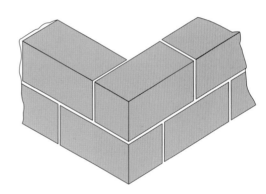

Figure 8.6 215 mm corner.

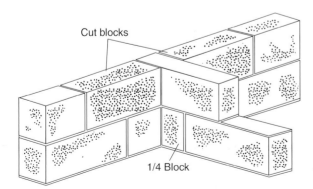

Cut blocks

1/4 Block

Figure 8.7 Block central to junction.

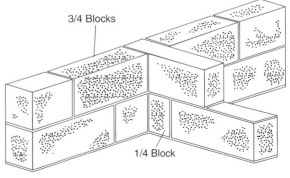

3/4 Blocks

1/4 Block

Figure 8.8 Block off centre.

Block Indents

Indents in blockwork can be formed to allow for additional walls to be tied in at a later stage of construction. The indenting of blockwork may be necessary to provide ease of access for the loading out of materials or free passage for operatives.

Bonding to an Indent

The bonding of the blockwork is straightforward enough. In the illustrations shown the indents are formed to coincide with the second course of blocks. There are three methods shown of bonding this length of walling. There is little to choose between the three examples, but the experienced bricklayer should always be aware of the bond when tying in to indents. In Figure 8.9 the tie blocks bond without any undue cuts to the blocks.

In Figure 8.10, because the courses have been reversed, a small cut has been introduced to the second course.

Figure 8.11 shows two cuts on the second course, which is a little wasteful of blocks.

Always think ahead. Ask yourself the question, how am I going to tie into the block or brickwork using a minimum of cut blocks thereby reducing time and wastage of materials?

The use of indents is not confined to the bonding of brick and block walls together. Indents are also used in the case of blockwork, for example if a bricklayer is building a block wall that has a junction, it may not be convenient to build the adjoining wall at the same time as the main wall. In this instance the bricklayer will leave indents for completion of the work at a later stage.

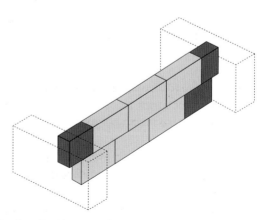

Figure 8.9   Bond without undue cuts.

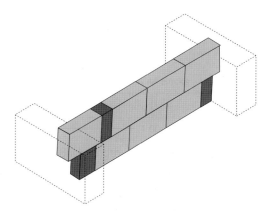

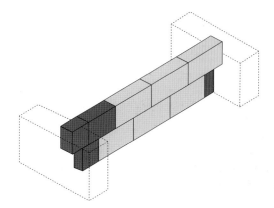

Figure 8.10 Small cut on second course.

Figure 8.11 Two cuts on second course.

Tool Skills

You are just beginning the Diploma, and acquiring the basic skills in bricklaying will take time to develop. What you are looking to achieve is dexterity, the skilful use of the hands. This has to be coupled with good hand–eye coordination.

To master the art of laying blocks and bricks, you will need to observe the actions of your tutor closely. This will be followed by regular practice of these actions: over time you will obtain the skills required.

The correct techniques are vital if full joints are to be achieved, the face of the blocks and bricks kept clean, the bricks laid at a productive rate and accuracy achieved.

Industrial Standards

All the work carried out during the Diploma will be marked to the Industrial Standards. These are standards of workmanship set by the construction industry, which are required on a building site from a competent craftsperson.

In the early part of your training you may find it difficult to achieve the required tolerances within the Standards. Your aim should be to fully understand the Standards and show continual progress towards meeting them.

To help you achieve this, self-assessment is used. On completion of a model, you will mark your own work to the Industrial Standards. This will then be checked by your tutor.

Finally, the industrial standards should be studied prior to commencement of work and strictly adhered to.

Good Practice When Laying Blocks

Solid and cellular blocks should be laid on a full bed of mortar and vertical joints substantially filled. Hollow blocks should be the same. Do not wet the blocks before laying them. Where necessary, adjust the consistency of the mortar to suit the suction of the blocks.

When laying facing blocks, select the blocks from more than one pack as work proceeds to reduce the risk of banding or patchiness of colour in the finished walling.

The following good working practices should be followed when laying blocks:

- Blocks should not be laid when the temperature is at or below three degrees centigrade and falling unless it is at least one degree centigrade and rising.
- High standards of workmanship should be followed at all times.
- All bed and perpendicular (perpend) joints should be fully filled with mortar.
- Cavities should be kept clear from mortar droppings and other debris.
- Partially completed and new work should be protected at all times from bad weather.
- Movement should be controlled by the inclusion of movement joints and/or bed joint reinforcement appropriately positioned.
- To maintain a satisfactory appearance in facing work, wall dimensions should be based on coordinating block sizes.
- For facing work, a sample panel should be built to enable specifications and standards of workmanship to be agreed before construction commences.
- Special blocks should be used for lintels, sills and closing cavities.

Recommended Heights of Walling

Lift heights will be affected by block thickness, weight, type of block, wall type and mortar mix used. Weather conditions will also affect lift heights; they may need to be restricted in bad weather. Generally, lift heights should be restricted to seven full block courses, approximately 1575mm in a working day.

Energy Conservation

Changes to energy conservation requirements for buildings mean that areas where thermal bridging may occur should be considered carefully and taken into account in the overall design of the building.

Note

The use of bricks and other dense materials for the closing of cavities is now not permitted by the Robust Details Document (this is used in conjunction with the Building Regulations for the specification of materials). The reason being is that the use of aerated blocks enables the U-value of the wall construction to be maintained up to the door and window reveals and at roof level.

Words and Meanings

Thermal Bridge – A region within a building element, such as a mortar joint or a lintel, where the conduction of heat is higher compared with other parts of the building element.

U-Value – The rate of heat loss, in British thermal units per hour, through 300mm square of a surface (wall, roof, door, window or other building surface) when the difference between the air temperature on either side is one degree Fahrenheit. The U-value is 1 divided by the R-value.

R-value – This is a measure of the ability of a piece of material to resist the passage of heat through it. It is equal to the thickness of the material in metres divided by its thermal conductivity.

Resources for Blockwork

Damp Proof Course

There are two main types of DPC material: flexible and rigid.

- **Flexible:** the most common types (bitumen polymer, pitch polymer or polythene) are supplied in a range of widths including 110, 220 and 300 mm.
- **Rigid:** Two courses of engineering bricks bedded in mortar give stability in free-standing walls. Rigid damp proof courses are suitable to resist rising damp but not the downward flow of water.

The one most commonly used for blockwork is the flexible type.

Insulation

Insulation is essential in order to meet Part L of the Building Regulations. It sits easily within the cavity and does not need clips if the cavity is fully filled. Insulation materials are thermally efficient and are relatively cheap.

Wall Ties

Wall ties are mainly produced from stainless steel as this material does not corrode over time. Wall ties resist lateral forces on the wall and increase the width of the cavity wall, making it more stable.

Mortar

A high-durability mortar mix of one part cement, half a part lime and four parts sand is normally suitable for all levels in the wall. Ensure you fully fill in all bed and perpend joints.

> *Remember* The mortar is exposed just as much as the block.

Blocks

There is a wide variety in the size and type of block used within the construction industry. The two most common types are as follows:

Lightweight Aircrete Blocks

These are manufactured in various widths of between 75, 100, 150 and 200 mm. They are designed for use in lightweight internal partitions and the inner leaves of cavity walls.

Aircrete is a lightweight, load-bearing and thermally insulating building material most commonly available in block format, but also as reinforced units.

Dense Concrete Blocks

These are manufactured in widths of 75, 100 and 150 mm. They are mainly used for load-bearing and exposed work on industrial and agricultural buildings. Aggregate concrete blocks are strong, durable, ideal for all wet finishes and for dry lining and are easy to lay and position on mortar.

Concrete blocks are manufactured under four headings:

Solid blocks

These blocks are widely used in industrial and agricultural buildings, providing a hard-wearing load-bearing structure. Solid blocks contain no formed cavities.

Cellular blocks

Used mainly for lightweight partitions. They are manufactured in widths of 75, 100 and 150 mm. Cellular blocks contain one or more formed cavities that do not penetrate the block.

Hollow blocks

These, mainly concrete, blocks are used for load-bearing and hard-wearing walls in industrial and agricultural buildings. Hollow blocks contain one or more formed cavities that fully penetrate the block.

Reveal blocks

These blocks, the majority of which are of the aerated type, are used to close cavities at door and window openings. Closing cavities with aerated blocks provides a number of benefits, both at the design stage and during construction.

Specifications

Specifications are written instructions that tell you exactly what materials to use for the construction work.

An example of laying blocks to a specification would be as follows.

Cavity return wall

Build the cavity return wall in dense concrete blocks as shown in the working drawing with both ends sealed with proprietary type closers. Joints to be half round to front face and rear face left flush from the trowel. Mortar to be a sand and cement mix with a ratio of 4:1.

Check all information provided on the working drawing and specification before commencing work.

Suitability of Resources

Your company's materials and equipment may come from an outside supplier, a plant hire firm or your own firm's stores. Whoever supplies them, you must be satisfied that they are exactly what were ordered and that their condition is satisfactory.

Handling Blocks Safely

Guidance by the Construction Industry Advisory Committee (CONIAC) covers the safe handling of building blocks of whatever type.

Hazards

The main hazards associated with laying blocks are:

- heavy loads and poor posture: excessive stress and strain causing injury to muscles and tendons, particularly where handling involves bending, twisting or other difficult postures;
- slips, trips and falls: including damage done by 'dropping blocks';
- sharp edges: cuts and abrasions to the skin;
- skin hazards: dermatitis, burns and similar conditions caused by contact with mortar.

Manual Handling

Careful consideration of the bricklayer's working area can contribute significantly to safe working and the reduction of hazards. Points to consider include:

- Move blocks in packs by mechanical means whenever possible.
- Load blocks out to above knee height.
- Ensure that normal PPE appropriate to construction sites is both provided and used.
- Ensure that appropriate eye protection equipment and dust suppression or extraction measures are provided when mechanically cutting or chasing blocks.

Inaccuracies in information

If you come across an error in an information source, for example a wrong measurement on a working drawing or the wrong type of block or brick in a specification, you must report it either verbally or in writing in the form of a memo to your tutor or supervisor. This will save your company or organisation the time and money that will be necessary to put the error right at a later stage.

Handling and Laying Blocks

Risk of injury is largely governed by the weight of the block: the heavier the block, the higher the risk of injury.

There is a high risk of injury in single handling and repetitive handling of blocks heavier than 20 kg.

If single-person handling is required, blocks of either 20 kg or lighter should be specified and used, or other precautions should be taken to reduce the risk:

- Stack blocks close to where they will be used.
- For blocks weighing 20 kg or more, employ mechanical lifting and handling.
- Handle blocks close to the body.
- Avoid over-reaching or twisting when handling and laying blocks.
- Only handle and lay blocks up to shoulder height.
- Use scaffolding if work is above shoulder height.

Training

Bricklayers should be given information and training on the safe systems of work and the procedures to follow to ensure the safe handling and laying of blocks.

Setting Out the Workstation

The positioning of materials and components prior to laying blocks is called setting out the workstation.

The materials must be placed to ensure economy of movement for the bricklayer with everything within easy reach.

To begin with, an approximate calculation is made of the materials needed. The blocks required are then spread out evenly along the length of the wall in neat bonded stacks. Mortar boards should be placed evenly between these stacks packed up on bricks or blocks. This helps to keep the work area tidy and reduces the distance the bricklayer has to travel for mortar.

The stacks and mortar boards should be approximately 600 mm from the face of the wall to give adequate working space (Figure 8.12).

Before mortar boards are loaded out with mortar, they should be wetted to ensure they do not absorb the water from the mortar mix.

Personal Protective Equipment

When handling blocks, the normal personal protective equipment (PPE) needed for use on construction sites should be provided by employers and worn by operatives. PPE should include safety helmet, safety footwear with protective toecaps and suitable gloves.

Figure 8.12 Example of a tidy workstation.

Cutting Blocks

Dense blocks are cut using the same process as bricks. The bolster chisel and lump hammer are most commonly used, again cutting on all four sides. However, lightweight blocks can be cut using a masonry saw. They can be cut with a bolster chisel and hammer but the material absorbs the shock and does not crack as easily as the dense block or brick.

On site, bricklayers will often use an electric angle grinder with a stone-cutting disc or tungsten-tipped steel blade to cut both dense and lightweight blocks.

Building Blockwork Corners

Laying concrete and lightweight insulation blocks is one of the most common tasks performed by a bricklayer.

Setting Out

Remove any debris and sweep clean a flat concrete floor surface approximately 3m × 2m. Draw a chalk line on the floor approximately 2m long using a straight edge. Lay out a line of blocks with allowances for 10mm joints along the chalk line to a length of 1.77m. Use a builder's square to form a right angle and repeat above, ensuring you allow for a quarter block at the quoin header to form the bond.

Note
For building corners, each quoin block when laid is checked for:
- gauge
- level in each direction
- plumb at each face
- face plane alignment.

133

Building the Corner

As stated, corner positions are marked on a chalk line drawn on the workshop floor. From each quoin block, further blocks are laid to raise the corner.

Stage one

Bed a quoin stretcher. Level the block both ways. Check for the correct gauge. Plumb both faces at the corner. Lay two blocks from the quoin stretcher and level the blocks from the corner.

Stage two

Plumb the end block (Figure 8.13). Use the spirit level as a straight edge to check the face alignment. The spirit level is held at a slight angle to touch the overall length of the corner.

Repeat the full sequence to lay and check each course until the corner is raised to its full height.

Stage three

On completion of the corner, the raking back is checked for alignment on each face (Figure 8.14).

As blocks are laid, surplus mortar will be squeezed out of the joints. To ensure that the face of the blocks are kept clean, the trowel blade is kept at an open obtuse angle to the wall face to remove the surplus mortar.

The surplus mortar on the rear of the block should also be removed (Figure 8.15).

Note

Perpend joints in alternate courses of blockwork should be vertical.

(a)

(b)

Figure 8.13   (a) Levelling and (b) Plumbing.

Figure 8.14 Checking racking back.

Figure 8.15 Removing surplus mortar.

Building Blockwork Walls

Laying concrete and lightweight insulation blocks is one of the most common tasks performed by a bricklayer.

Setting Out

Remove any debris and sweep clean a flat concrete floor surface approximately 3 m × 2 m. Draw a chalk line on the floor approximately 2 m long using a straight edge. Lay out a line of blocks with allowances for 10 mm joints along the chalk line to a length of 1.8 m. Build corners as previously described.

Building the Wall

The most effective method of laying lightweight or concrete blocks is to use a bricklayer's line as a guide.

Laying blocks to a line requires the use of many repetitive activities such as:

- spreading mortar
- placing blocks
- placing cross-joints.

Stage one

Now let us study the procedures that must be followed to lay blocks to a line.

- Fix the bricklayer's line from corner to corner using the method shown by your tutor.

- Place the blocks along the line leaving enough space for spreading the mortar.
- Spread the mortar for bedding the first course (Figure 8.16).

Stage two

- Starting from the corner, lay each block to the line, gauging the cross-joints (Figure 8.17).
- The top edge of the block should be level with the top of the line (Figure 8.18).
- The clearance between the blocks and the line should be approximately 2 mm (Figure 8.19).

Figure 8.16 Spreading mortar.

Figure 8.17 Laying blocks.

Figure 8.18 Block level with line.

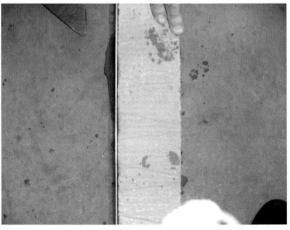

Figure 8.19 Clearance between block and line.

Figure 8.20   Filling cross-joints.

- Fill the cross-joints solidly (Figure 8.20).

Stage three

- For each subsequent course of blocks, fix the line to the corners. Make sure that the line is fixed to the same course height at each corner.
- Place the blocks to be laid along the wall and spread mortar for the bed joint along the top of the previous course.
- Lay the blocks to the line, taking care to regulate cross-joints and check the positioning of the top edge of the block with the line.
- Fill all cross-joints and clear away all excess mortar.

> **Note**
>
> The joint finish must be uniform and extend to each arris (i.e. the edge of the brick or block) as appropriate to the joint profile.

Joint Finishes to Blockwork

Jointing can be formed with a pointing trowel or an appropriate jointing tool as the work proceeds. Form the cross-joints first.

Form the bed joints next. The bed joint finish must be uninterrupted in its length.

Lightly brush the wall to remove mortar crumbs from the brick or block arrises. Use the side of the brush only.

Weather conditions, type of brick or block and mortar composition can all vary. This can affect the time when the mortar joints become suitable for the jointing process to begin. When building face blockwork, test the joints to determine the optimum time to form the joint finish.

The final appearance and weather-resisting qualities of the wall depend on the joint finish.

> **Note**
>
> At a corner, form the joint finish from the corner in each direction. This ensures a full joint at the corner.

> **Remember** Your performance will be evaluated as not yet competent if you fail to meet the industrial standards.

Workshop Drawing

A workshop drawing is a way of showing what a blockwork model will look like and how it needs to be built. It is not a picture, but a tool for working with.

A workshop drawing shows different views, or elevations, of a block-work model that you need to build to complete the practical aspects of the Diploma.

Measurements

On the drawing, you will find all the measurements you need: length of the wall, size of doors and windows, the height of the damp proof course and so on. You will also find other information about where air bricks go, what type of lintels you need, type of bond and so on.

Everything on the workshop drawing is made smaller by using scale. That means the proportions are right but it is all much smaller – a miniature version of the real thing.

Scales Commonly Used

It is impossible to draw buildings, plots of land and workshop models to their full size. However, you can draw them small enough to fit on a piece of paper. This is known as drawing to scale.

The main scales commonly applied to workshop drawings are as follows:

- 1:10 – 10 times smaller than the model
- 1:20 – 20 times smaller than the model
- 1:50 – 50 times smaller than the model.

A scale rule is often used to measure workshop drawings.

Following and Confirming Instructions

Listen Carefully

You will need to listen to and read given instructions at college and in the workplace. It is important to understand that these instructions are not always given in the order that the jobs need to be carried out.

Following verbal instructions so that they can be carried out in the intended order requires good listening skills. Often, you will have to listen carefully to pick out exactly what you are being asked to do from things that people say.

This is why it is important to check that you understand given instructions, be they verbal or written: failure to do so could mean you carry out the wrong activity or do nothing at all.

Confirming Instructions

If you have been given a task to carry out, for instance building a wall either in writing or verbally, always confirm these instructions with your tutor or supervisor to make sure you are carrying out the task correctly.

Check Your Sources

Always check that information sources, such as working drawings or specifications, comply with good practice. Your tutor or supervisor will have a checklist for you to follow and will explain what good practice is in relation to various activities.

Recording Discrepancies

If you notice a discrepancy in information, for example the wrong measurements on a working drawing at college or work, make a written note of it, so you don't forget it. Then at a suitable time you can inform your tutor or supervisor and they can decide what action to take.

> *Remember* If you are unsure of something, always ask for clarification.

> *Remember* You are learning, so no one will mind you double-checking work to be carried out.

Information

For further information about blocks and other associated products, visit the websites of the better-known manufacturers and suppliers or the Concrete Block Association (CBA): www.cba-blocks.org.uk.

The CBA is the trade body that represents manufacturers of aggregate concrete building blocks in Great Britain. The CBA represents an industry of some 50 manufacturers producing around 60 million square metres of concrete blocks per year from over 100 block plants nationwide.

Information on the Thin Joint Blockwork System

For information about the thin joint blockwork system, visit the websites of the better-known manufacturers and suppliers.

Quick Quiz

1. Name two types of block.

2. What is the minimum overlap when bonding blockwork?

3. Where would you locate broken bond in a block wall?

4. Below what temperature should blocks not be laid? What is the exception to this rule?

5. What is the recommended lift height for blockwork in a working day?

CHAPTER 9

Numeracy Skills

THIS CHAPTER RELATES TO UNIT CC1002K AND UNIT CC1002S

Calculating Quantities of Materials

Calculating quantities of materials is an important aspect of building, particularly for bricklayers. This chapter looks at methods of calculating the numbers of bricks and blocks in a wall. Working drawings do not tell you how many bricks and blocks you need to build a wall. You work this out using the measurements shown on the working drawing.

Perimeter and Linear Measurement

The perimeter of a shape is easy to work out. It is just the distance all the way round the edge, also known as the linear measurement. If the shape has straight sides, add up the lengths of all the sides. These may be given or you may need to measure carefully along each of the sides using a ruler.

Calculating Quantities of Blocks

To calculate how many blocks there are in one square metre ($1\,m^2$) of walling, we use the following method.

Stage one

Divide a 1m horizontal length (1000mm) by the length of a standard block and mortar joint. This will tell us how many blocks there are in the length. Therefore:

$$1\text{m divided by } 0.450\text{ m} = 2.22\text{ blocks}$$

Stage two

Divide a 1m vertical length (1000mm) by the thickness of a standard block and mortar joint. This will tell us how many blocks there are in the height. Therefore:

1m divided by 0.225 m = 4.44 blocks

Stage three

Multiply the number of blocks that go into the length of the wall by the number of blocks that go into the height of the wall. This will give us the number of blocks included in 1 m² of walling. Therefore:

$$2.22 \times 4.44 = 9.87 \text{ blocks}$$

We then round the figure up to the next whole number. Therefore, there are approximately 10 blocks in 1 m² of walling.

Calculating the Number of Bricks

We use the same method for calculating the number of bricks in a square metre of walling.

To work out how many bricks there are in 1 m² of brickwork, we use the following method.

Length of a brick plus mortar joint = 215 mm + 10 mm = 225 mm. Thickness of a brick plus mortar joint = 65 mm + 10 mm = 75 mm.

Stage one

Divide the length of a standard brick and mortar joint into 1 m (1000 mm) to find out how many laid bricks make a length of 1 m (Figure 9.2).

1000 divided by 225 = 4.44 (bricks)

Stage two

Divide the height of a standard brick and mortar joint into 1 m (1000 mm) to find out how many laid bricks make a height of 1 m (Figure 9.3).

1000 divided by 75 = 13.33 (bricks)

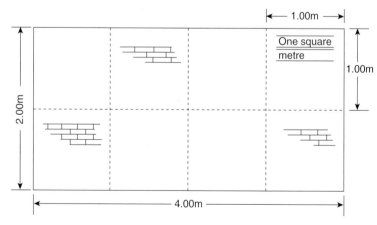

Figure 9.1 Measuring the area of a 4 m × 2 m wall.

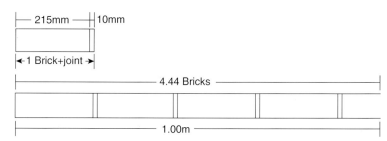

Figure 9.2 Brick length divided into 1 m.

Stage three

Multiply the number of bricks that make a length of 1 m by the number of bricks that make a height of 1 m.

This gives you the number of bricks that make an area of 1 m².

$$4.44 \times 13.33 = 59.18 \text{ (bricks)}$$

Round this up to the next whole brick: there are approximately 60 bricks in 1 m² of brickwork.

Figure 9.4 shows what 1 m² of brick or blockwork looks like.

So for our wall above which is built in half brick walling, we would make the following calculation:

Bricks per metre squared = 60

Area of the wall = 4 m × 2 m = 8 m²

Bricks required for 8 m² of walling: 60 × 8 = 480 bricks

How to do a Brick Count

To work out how many bricks to order, you need to:

- Find out how many bricks there are in 1 m² of brickwork using the method shown.
- Work out how many square metres of brickwork you are going to build. Work this out from the measurements on the working drawing.
- Multiply the number of bricks in 1 m² by the number of square metres of brickwork you want to build.

Block Quantities

The calculation of brick and blockwork quantities is something that a bricklayer will constantly encounter. The method of calculation is fairly

> **Remember** You need to convert 1 m to 1000 mm before you can do your calculation.

> **Note**
>
> There are 60 bricks in 1 m² of half brick walling and 120 bricks in 1 m² of one brick walling. There are 10 blocks in 1 m² of walling.

> **Tip**
>
> The method for calculating the number of bricks you require can be used no matter what the size of the wall is, and you can use it to calculate numbers of blocks, instead of bricks, as well.

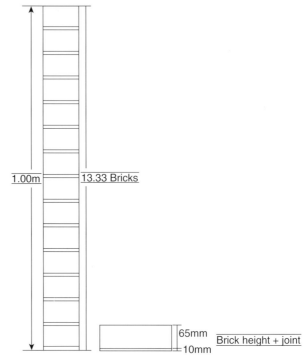

1.00m   13.33 Bricks

65mm   Brick height + joint
10mm

Figure 9.3   Brick height divided into 1 m.

simple, as long as you remember to follow a logical sequence. To calculate the area of blockwork, we follow the following sequence of actions:

$$Area = length \times height$$

Mortar Requirements

Rules

1 kg of mortar will lay one brick.

2 kg of mortar will lay one block.

Therefore, for one square metre of half brick walling you would need 60 kg of mortar (1 kg × 60 bricks = 60 kg) and for one square metre of one brick walling you would need 120 kg of mortar (1 kg × 120 bricks = 120 kg).

For one square metre of blockwork you would need 20 kg of mortar (2 kg × 10 blocks = 20 kg).

So for our wall above we would need 480 kg of mortar (1 kg × 480 bricks = 480 kg).

Areas of Basic Shapes

Calculating the Area of a Rectangle

The area of a shape is the amount of surface that it covers, like a wall or floor. You need to know the areas of walls so you can work out how many bricks and blocks you need for the job.

Finding the area of a wall is easy if you know the length and width. The wall in Figure 9.1 is 4 m long and 2 m high.

As you can see from the illustration, this gives you eight 1-metre by 1-metre squares. So the area is eight square metres ($8 m^2$). The quick way to work this out is to measure the length and the width/height and multiply them together. For the wall in Figure 9.1, this will look like this:

$$4 m \times 2 m = 8 m^2$$

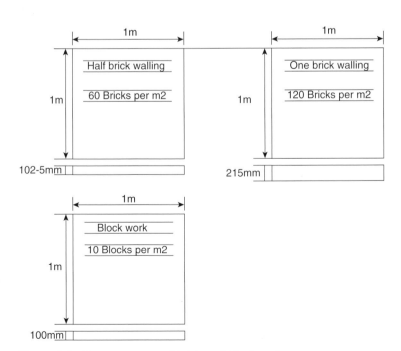

Figure 9.4 Square metres of brick and blockwork.

The formula looks like this:

$$\text{Length} \times \text{Width} = \text{Area}$$

Establishing the amount of materials required to complete a job is essential if the job is to be completed on schedule and within budget. This means calculating the area of a wall from a working drawing in order to estimate the number of bricks needed to build that wall.

Dimensions on drawings are normally shown in millimetres but area needs to be measured in square metres (m^2). It is easier to convert the dimensions on the drawing into metres before you calculate the area.

To convert a dimension from millimetres to metres you have to divide it by 1000.

This is because there are 1000 mm in 1 m.

In Figure 9.5 2000 = 2 m because 2000 divided by 1000 = 2 and 6055 = 6.055 m because 6055 divided by 1000 = 6.055.

So the area of this wall is:

$$
\begin{aligned}
A &= L \times W \\
&= 6.055\,\text{m} \times 2 \\
&= 12.11\,\text{m}^2
\end{aligned}
$$

Calculating the Area of a Triangle

Not all walls will be simple rectangles. You will have to work out the areas of walls with gable ends.

The gable end of a building is usually a triangle. To work out the area of a gable end, you need to know how to find the area of a triangle (Figure 9.6).

To calculate the area (A) of a triangle:

Multiply the base (b) of the triangle by the perpendicular height (h). Divide this by 2. The formula looks like this:

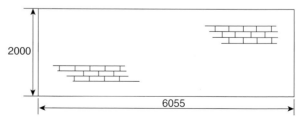

Figure 9.5   Millimetres to metres.

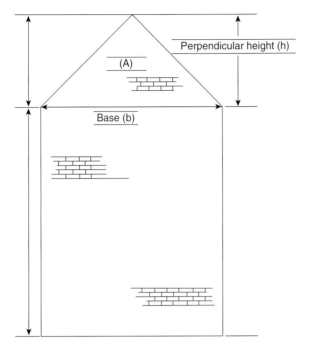

Figure 9.6 The gable end of a house.

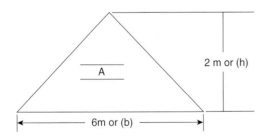

Figure 9.7 Calculating the area of a triangle.

$$A = b \times h \text{ divided by } 2$$

- The perpendicular height (h) of this triangle is 2000 mm or 2 m.
- The base (b) of this triangle is 6000 mm or 6 m.
- So, the area of the gable end is:

$$A = b \times h \text{ divided by } 2$$
$$= 6 \text{ m} \times 2 \text{ m divided by } 2 = 6 \text{ m}^2.$$

Calculating the Total Area of an End Wall

Now that you know how to calculate the area of a gable end (Figure 9.8), you can work out the total area of the wall. The total area of the wall is equal to the area of the triangle plus the area of the rectangle:

$$\text{Area of triangle} = b \times h \text{ divided by 2}$$
$$= 6 \text{ m} \times 2 \text{ m divided by } 2 = 6 \text{ m}^2$$

$$\text{Area of rectangle} = L \times W$$
$$= 6 \text{ m} \times 8 \text{ m} = 48 \text{ m}^2$$

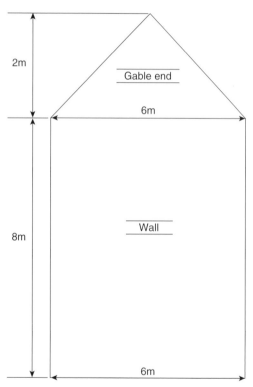

Figure 9.8 A gable end and wall: diagram.

$$\text{Total area of wall} = \text{Area of triangle} + \text{Area of rectangle}$$
$$= 6 \text{ m}^2 + 48 \text{ m}^2$$
$$= 54 \text{ m}^2.$$

Try this Out

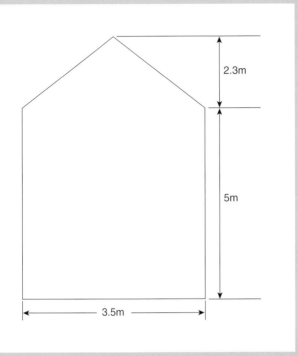

Use the formulas for finding the area of the triangle and rectangle to calculate the area of the gable end and wall illustrated in Figure 9.9.

Tip The total area of a wall with a gable end is equal to the area of the rectangle plus the area of the triangle.

Figure 9.9 A gable end and wall: measurements.

Calculating Volumes

Volume is the amount of space taken up by a three-dimensional (3-D), or solid, shape. Volume is measured in cube units, such as cubic centimetres (cm^3) and cubic metres (m^3).

The formula for the volume of a cuboid (Figure 9.10) is:

$$volume = length \times width \times height$$

The formula for the volume of a cube (Figure 9.11) is the same:

$$volume = length \times width \times height$$

Words and Meanings

Cuboid – A 3-D shape whose faces are all rectangles.

Words and Meanings

Cube – A 3-D shape whose faces are all squares.

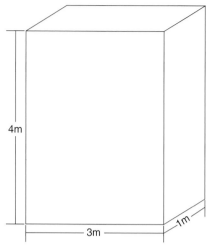

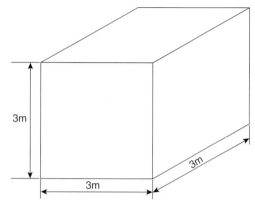

Figure 9.11   A cube.

Figure 9.10   A cuboid.

Areas of Circles

The area of a circle is expressed using the formula: $3.142 \times$ the radius squared, or Pi $\times$ radius squared.

Terms Used

The following terms are used when calculating the area of circles (Figure 9.12):

- Pi – a constant value number which is used in determining the area or circumference of a circle. The value of Pi is 3.142.
- Radius – a straight line running from the centre of a circle to the circumference.
- Diameter – This is a straight line which passes through the centre of the circle and terminates at the circumference.
- Circumference – a linear measurement of the outside line of a circle.
- Radius – The radius is the distance from the centre of the circle to the circumference. It is always half the length of the diameter.
- Chord – A chord is a straight line which touches the circumference at two points but does not pass through the centre.
- Arc – An arc is any section of the circumference.
- Normal – A normal is any straight line that starts at the centre and extends through the circumference.
- Circle – a round shape that is measured in square metres.

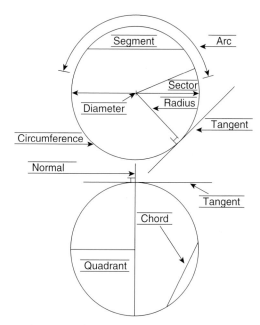

Figure 9.12 The parts of a circle.

- Tangent – A tangent is a straight line which touches the circumference at right angles to the normal.
- Sector – This is a portion of a circle which is contained between two radii and an arc.
- Quadrant – This is also a sector whose area is a quarter of a circle.
- Segment – This is the portion of a circle contained between an arc and a chord.
- Linear – a length given in metres.
- Surface area – total area enclosed by the circumference of the circle.

Examples

Calculate the area of a circle if the radius is 1.5 m.

$$\text{Area of circle} = \text{Pi} r^2$$

$$\text{Therefore} - 3.142 \times 1.5 \times 1.5 = 7.068 \text{ m}^2.$$

If a circle has a radius of 1.75 m, what is its surface area?

$$\text{Surface area of a circle} = \text{Pi} r^2$$

$$\text{Therefore} - 3.142 \times 1.75 \times 1.75 = 9.62 \text{ m}^2.$$

Estimating Material Quantities

Wastage

Sometimes bricks and blocks get damaged during delivery or cutting. Damaged bricks and blocks are referred to as wastage.

Before putting in your order for bricks and blocks, it is good practice to add on an extra amount for wastage. The usual allowance for wastage is 5%. There are several ways to calculate this.

Doing it in your head

Stage 1: Find 5%

Divide the number of blocks or bricks by 10 to get 10%, then divide your answer by 2 to find 5%.

Example: You need 120 blocks to build a wall.

$$10\% \text{ of } 120 = 120 \text{ divided by } 10 = 12$$
$$5\% \text{ is half of } 10\%$$
$$\text{Half of } 12 \text{ is } 6$$

You need to order 6 extra blocks.

Stage 2: Add on the wastage allowance

Add the amount you will allow for wastage to your original figure to work out how many blocks or bricks to order.

$$120 + 6 = 126$$

Order 126 blocks.

Question

Why does multiplying by 1.05 increase an amount by 5%?

This is how it works:

- 5% is the same as .0595 hundredths).
- Multiplying a number by 0.5 gives you 5 hundredths of that number.
- Multiplying a number by 1 gives you the original number.
- So multiplying a number by 1.05 gives you the original number plus 5 hundredths.

Using a calculator

It is always a good idea to use a calculator to carry out and check calculations. Using a calculator can give a quick and accurate solution in the workplace. However, you will need to understand the calculation first.

Metric measurements are recorded in decimals. It is usually better to convert measurements in millimetres to metres.

Example

> 457 mm is the same as 0.457 m. You key this in as 0.457.

The decimal point separates whole units from parts or fractions of a unit.

Numbers may be displayed to more decimal places than you need. Round the number to 1 or 2 decimal places, depending on the circumstances

The order in which you enter the numbers and symbols really does matter. Be clear about this before you carry out any calculations.

Metric and Imperial Measurements

The construction industry mainly uses metric units for measuring. However, you may come across some older measures called imperial units. Metric measurements are recorded in decimals. It is usually better to convert measurements in millimetres to metres.

Conversion Tables

Conversion tables are used to convert between metric and imperial units.

There are many types of conversion tables available. I have listed a few of the most useful conversions below.

Equivalent measures: metric

Length, mass (or weight) and capacity are all measured using different units.

Length

- 1 centimetre (cm) = 10 millimetres (mm).
- 1 metre (m) = 100 cm.
- 1 kilometre (km) = 1000 m.

Mass

- 1 kilogram (kg) = 1000 grams (g).
- 1 tonne = 1000 kg.

Capacity

- 1 litre (l) = 1000 millilitres (ml).
- 1 centilitre (cl) = 10 ml.

Converting between metric units

Converting between metric units is always about multiplying or dividing by 10, 100 or 1000.

Equivalent measures: Imperial

In the past, we used imperial measures. We still sometimes use pints, gallons, pounds, inches and feet.

Length

- 12 inches = 1 foot
- 2.5 cm = approximately 1 inch
- 30 cm = approximately 1 foot
- 3 feet = approximately 1 metre.

Mass

- 16 ounces = 1 pound (lb)
- 25 grams (g) = approximately 1 ounce
- 2.25 lb = approximately 1 kg.

Capacity

- 8 pints = 1 gallon
- 1.75 pints = approximately 1 litre
- 4.5 litres = approximately 1 gallon.

Information

For more information about numeracy, visit your local or college library and take out some books on numbers related to brickwork.

Quick Quiz

1. How many bricks are there in one square metre of one brick walling?

2. What is the formula for working out the area of a triangle?

3. How would you find the volume of a cube?

4. How many blocks are there in five square metres of walling?

5. How much mortar would you need to lay ten blocks?

CHAPTER 10

Bricklaying Skills

THIS CHAPTER RELATES TO UNIT CC1015K AND UNIT CC1015S.

Setting Out a Workstation

The positioning of material prior to laying bricks is called setting out the workstation. The bricks and mortar must be placed to ensure economy of movement for the bricklayer with everything within easy reach.

First, a rough calculation of the bricks or blocks required should be carried out and the total number of bricks or blocks spread evenly along the length of the wall in neat bonded stacks.

The stacks should be approximately 600 mm from the face of the wall to give adequate working space (Figure 10.1).

Brick stacks should be bonded in order to provide stability. Brick faces should be facing away from the mortar board to avoid staining. Mortar boards should be packed up on bricks as this helps to keep the work area tidy and reduces the distance the bricklayer has to bend down to pick up mortar.

Before the mortar boards are loaded up with mortar, they should be dampened to ensure they do not absorb the water from the mortar, which weakens the mortar mix and spoils the workability of the mortar.

Good housekeeping

Your tutor in the first week of the programme will show you the correct way to set up your workstation prior to you carrying out bricklaying activities.

Figure 10.1 Workstation set out on the boarding of an independent scaffold ready to commence face brickwork.

Try this Out

Set up workstation in the manner shown (Figure 10.2), to include:

- setting out for walling by chalk line
- stacking of bricks or blocks neatly spaced from each stack and proposed wall
- mortar board in between bricks or blocks and 'blocked up on bricks
- all necessary tools placed to one side of wall.

On completion, ask your tutor to assess the workstation to see if it meets the criteria.

Figure 10.2 Example of a workstation.

Building a Corner

Setting Out

Setting out the corners of a brick or block building is one of the first activities to be carried out on a building site. When setting out corners for face brick walls, where all the brickwork is exposed and not rendered with mortar, care must be taken to create a pleasing exterior wall appearance, if possible free of any cut bricks.

When setting out for corners, any bonding adjustments required should be made when initially setting out.

The first stage in setting out is to mark the corners. Plumb down from the profile line intersecting points and mark these points on the concrete strip foundation. These are your corner points.

Now, fix or mark lines between all corner points. Next, dry bond the first course along the fixed lines, gauging the size (width) of the cross joints using a 10 mm spacer as the bricks are laid out.

If the bonding does not work out as required by the specifications, adjust by increasing or decreasing the cross joints.

Now you are ready to establish the position of the return corner.

Method

From each quoin brick, further bricks are laid to raise the corner.

Each course is checked for level, plumb and face alignment from the quoin bricks.

Bed a quoin stretcher. Level both ways and check for correct gauge. Plumb both faces at the corner and lay two bricks from the quoin stretcher. Level bricks from the corner (Figure 10.3).

Plumb end brick. Use the spirit level as a straight edge to check the face alignment. The spirit level is held at a slight angle to touch the

Figure 10.3 Levelling.

Figure 10.5   Removing surplus mortar.

Figure 10.4   Plumbing.

overall length of the corner. Repeat the full sequence to lay and check each course until the corner is raised to its full height (Figure 10.4).

On completion of the corner, the raking back is checked for alignment on each face. As bricks are laid, surplus mortar will be squeezed out of the joints. To ensure that the face of the brickwork is kept clean, the trowel blade is kept at an open obtuse angle to the wall face to remove the surplus mortar.

The surplus mortar on the rear of the brick should also be removed (Figure 10.5).

Building a Straight Wall

Now let us proceed to set out and build the wall. Using a straight edge and pencil, mark the face line (Figure 10.6).

Set out the first course without mortar (dry bond); use a 10 mm wooden spacer to measure the mortar joints and mark them (Figure 10.7).

Assuming that corners have been built at each end of the wall, the bricks between may then be tested for alignment along the top and face with a straight edge (Figure 10.8) or, where the distance is too great for a straight edge to be conveniently handled, with line and corner blocks.

Figure 10.6   Marking out the face line.

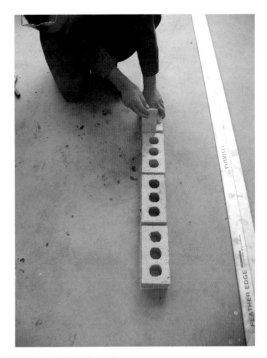

Figure 10.7   Dry bonding.

Figure 10.8   Checking alignment.

Figure 10.9 Working to the line.

In this case a line would be stretched from the top of one brick to the top of the brick at the other end, the line being held in position by the use of corner blocks. The bricks should then be laid so that their top edges just coincide with the inside of the line (Figure 10.9).

English and Flemish Bond Walls

English Bond

Make sure you have all the tools, equipment and materials you will need. Make sure your work area is clean and level. Now let us proceed to set out and build an English bond wall 890 mm long.

Using a straight edge and pencil, mark the face line of the wall. Set out the first course without any mortar (dry bonding). Use a 10 mm spacer to measure the width of the mortar joints and mark them on to the floor.

Lay No. 1 and No. 2 bricks in mortar – level and gauge them. Stretch a line between them and finish the course, or use your spirit level if it is long enough.

Lay No. 3 and No. 4 bricks in mortar and check the wall width (Figure 10.10).

Level and range them to brick No. 1 and brick No. 2. Use the line or level to complete the course (Figure 10.11).

> **Note**
>
> The width of a one-brick wall is equal to the length of one brick (215 mm). Check the width.

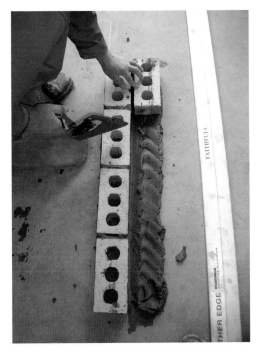

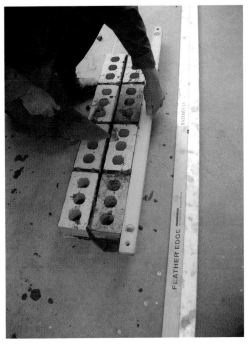

Figure 10.10   Laying first course of a one-brick wall in English bond.

Figure 10.11   Levelling and ranging first course of a one-brick wall in English bond.

Building the second course

First, lay the two end bricks as illustrated in Figure 10.12a. Gauge, level and plumb these two bricks. Fill in the remaining bricks to the course, plumbing and levelling as illustrated in Figure 10.12b.

Procedures to completion

Alternate the first and second course pattern until the wall is ten courses in height. Plumb, level and gauge each course laid.

Check the mortar joints to find if they are ready to be tooled. When ready, tool all joints smooth and straight.

Continue to practise building this wall until you feel confident that your workmanship measures up to the industrial standards. Then ask your tutor to assess the wall.

Flemish Bond

Now apply this method to a Flemish bond wall, that is to a 215 mm wall 890 mm long.

Make sure you have all the tools, equipment and materials you will need and make sure your work area is clean and tidy.

(a)   (b)

Figure 10.12   Completing second course of a one-brick wall in English bond.

Resources

Bricklayers and other building workers must be able to identify the different types of bricks and blocks and their uses. Clay and concrete bricks and lightweight and dense concrete blocks are used to build houses, apartments, schools, shops and offices.

The most frequently used building unit is the brick. A brick is a block of material used for building or paving purposes. Bricks can be made from clay, clay mixtures or cement and sand and are usually baked in a kiln. The dimensions of bricks are 215 mm × 102.5 mm × 65 mm.

Brick Classifications

Bricks may be classified according to their uses as follows, bearing in mind that it is sometimes possible for a brick to come under more than one heading:

- facing bricks
- common bricks
- engineering bricks
- special bricks.

Words and Meanings

Facing – Covers a very wide range of bricks since it includes all those used for exterior and interior walls that are to be left as finished work.

Facing bricks

Facing bricks are intended to provide an attractive appearance. They are available in a range of brick types, colours and textures. Some may not be suitable in positions of extreme exposure. Some facing bricks have engineering properties.

Common bricks

Common bricks are suitable for general building work not chosen for its appearance. These are bricks for ordinary work that is not exposed to view, for example walls that are to be plastered or built underground.

Engineering bricks

Engineering bricks are hard burnt bricks that are very dense. They have a minimum compressive strength and minimum water absorption. They are not chosen for their appearance. There is no requirement for colour.

These are bricks suitable for ground works, manholes and sewers, retaining walls or as a ground level damp proof course to free-standing walls and situations where high strength and low water absorption are the most important factors.

Special bricks

A wide variety of bricks are available in special shapes or sizes, to blend or contrast with most facing bricks.

Squints are an example. They are manufactured to special shapes that enable the bricklayer to build angled corners at 45 degrees or 60 degrees. They are used to reduce the thickness of a wall and still maintain the face texture of the wall or remove the sharp corners from a brick wall or pier.

Block Classifications

The two most common types of blocks used in the construction industry are insulation blocks and concrete blocks.

Lightweight insulation blocks

Aircrete is a lightweight, load-bearing and thermally insulating building material most commonly available in block format. The blocks are light in weight, easy to work and have excellent load-bearing capabilities.

Concrete blocks

Concrete blocks can be used for load-bearing walls if they meet standards for bearing wall units. They are usually very heavy.

Concrete blocks are frequently used for constructing walls as they are larger and more economical to use than bricks. The smallest block is equal to six ordinary bricks.

Ask your tutor to show you some samples of the bricks and blocks commonly used in your locality.

Cutting Bricks

Before cutting a brick, the position of the cut must be measured off and a square mark pencilled on with the aid of a try square. A specially prepared gauge may be used for marking off the lengths of closers, bats and three-quarter bricks (Figure 10.13).

After marking, the brick should be placed on a solid bed with the portion nearest to you and the bolster chisel then held firmly on the brick, the top leaning slightly towards the waste part of the brick (Figure 10.14).

The bolster chisel should then be struck lightly with the club hammer. This operation, which is known as scoring the brick, is repeated on all four sides. The brick or block is then placed in its original position and finally cut through with a sharp blow, the force of the blow varying with the type of brick or block to be cut. Note the illustration, in which the hammer is held at the end of the handle; in this way the maximum force is obtained with the minimum of effort (Figure 10.15).

In many instances, this would be all the cutting required, unless there are portions of the brick or block projecting beyond the required face or end remained; in which case, they should be removed with a scutch.

For convenience and ease of working, it is an advantage if the brick or block is held down on a bench or similar raised surface when carrying out this final trimming, but a working platform seldom offers such conditions; therefore the brick must be held firmly in the hand as shown in Figure 10.16. When trimming a brick or block, the edges should be dressed by working the blade of the scutch diagonally along them, leaving the centre until last.

Figure 10.13   Brick marking gauge.

Figure 10.14   Preparation for cutting.

Figure 10.15 Cutting brick.

Figure 10.16 Trimming brick.

Figure 10.17 Using a bevel.

Bevel

A bevel is used for marking off an angle in the manner described for a square. The blade is pivoted at the top of the stock and fixed to enclose the required angle by fastening with a set screw (Figure 10.17).

Risk Assessment

Bricklayers and all construction site workers should assess personally the risk of working in particular settings or on particular jobs. At Diploma level, you will be recording risk assessment information, so you should be able to read and understand it. This section covers the principles of risk assessment so that you can assess information and take responsibility for your own and others' health and safety.

Employers carry out formal risk assessment and write down safe working methods or method statements. You should also carry out a risk assessment for yourself every time you start a job in the workshop.

There are three main stages to risk assessment:

Stage One

Look for the hazards. Look for things that can cause harm or areas where accidents are more likely to happen.

Stage Two

Think about who is at risk from this hazard. Is it everyone or only workers doing one particular job? Are members of the public at risk?

Stage Three

What can be done to reduce the risk? Is there a safe method of working that will help to prevent accidents? Who is responsible for carrying out these safety precautions?

Safety Rules

Building sites are dangerous places. Risk assessment involves identifying hazards and thinking of safe solutions.

You have a responsibility to conduct all your activities in a safe manner. It is against the law for you to endanger yourself or others by your actions or omissions. A copy of the rules will be given to you by your tutor or supervisor.

Your duty to yourself and others covers:

- Before, during and after carrying out any work, the workplace must be clean and tidy to prevent tripping and falling.
- Organise the work and set up tools and equipment correctly.
- Check all equipment is in good working order. Setting up to carry out your work effectively and efficiently can save time and risk of injury to yourself and others.
- Always wear the appropriate personal protective equipment (PPE).

Words and Meanings

Hazard – A danger or risk.
Hazardous – Dangerous, harmful.
Risk – Likelihood or chance that harm, illness or damage will occur and the degree of harm (how many people may be affected and how badly).
Risk Assessment – Mainly carried out by an employer to identify risks to his or her employees (and others) and to decide what is necessary to control these risks to the standards required under the law.

You must also be aware of emergency procedures in the event of an accident (also have a look at the relevant sections in Chapter 1); this includes knowing:

- who the first-aider is
- where the first-aid box is located
- how to fill in the accident book
- report accidents to the person in charge.

In the event of fire, carry out the correct emergency procedures for calling for help and evacuating the building (again, have a look at the sections in Chapter 1).

These rules will be read to you and explained by your tutor at the beginning of term.

Remember Safety is everyone's responsibility.

Checking Brickwork for Accuracy

It would be impossible for every single joint in a wall to be the same size and for every single course to be perfectly level and vertical (plumb). However, it is important that all brickwork be completed to industrial standards. These require that the bricklayer lay bricks and blocks accurately. The tolerance can be as little as plus or minus 3mm.

Tolerances tell you what is acceptable and what is not acceptable.

Tolerances are written using the symbols + (plus) and − (minus) together.

The symbol is always followed by a measurement that tells you how accurate your work needs to be. For example, a standard vertical joint is 10mm thick with a tolerance of +/−3mm (plus or minus 3mm). This means that vertical joints are acceptable as long as they measure no more than 3mm wider or no more than 3mm narrower than 10mm. In Figure 10.18 all of the vertical joints are acceptable. But in Figure 10.19, neither of these vertical joints is acceptable.

The amount of tolerance that you are allowed is usually very small.

A wall 3m high should be built to gauge +/−5mm in 3m height and with regular joint thickness.

This means that you must use a gauge staff to check that the joint for each course is a regular thickness. A tolerance is allowed for slight variations in joint thickness.

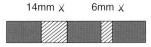

Figure 10.19 Unacceptable vertical joints.

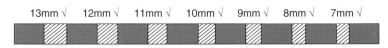

Figure 10.18 Acceptable vertical joints.

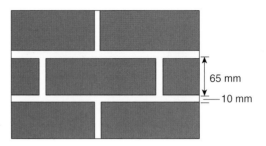

Figure 10.20 Gauge.

Words and Meanings

Gauge – The name given to the thickness of a brick plus the horizontal joint. In Figure 10.20 the gauge is 65 mm + 10 mm = 75 mm.

Your brickwork will be acceptable as long as it measures no higher than 3005 mm and no lower than 2995 mm after the required number of courses have been laid.

If built to gauge:

4 courses will measure 300 mm high (4 × 75 mm)

40 courses will measure 3000 mm (or 3 m) high (40 × 75 mm)

Understanding tolerances is important to your job. Tolerances tell you how accurate your work needs to be; they also tell you the largest difference from the standard amount that is still acceptable.

Specifications

During your workshop training, you will need to work to the required specification as described below.

Straight Wall

Build wall in stretcher bond with all joints flush from the trowel. Check the information provided on the drawing and project marking sheet before commencing.

Setting out

In a suitable work area, mark lines on the floor for setting out the model.

Set out the work area with the mortar boards and sufficient bricks to build the half-brick wall as shown in the drawing given to you by your tutor.

The drawing indicates the essential plumbing and ranging points that will assist you in building the wall accurately.

All the joints should be 10 mm thick, finished flush from the trowel and the surface of the brickwork clean and free from smudging.

The face planes of the wall should be plumb and level.

On completion of the model, your tutor will assess your work using the project marking sheet.

Following Instructions

You will need to listen to spoken instructions at college and in the workplace. It is important to understand that these instructions are not always given in the order that the jobs need to be done. Following spoken instructions so that they can be carried out in the intended order requires good listening skills. Often you will have to listen carefully to pick out exactly what you are being asked to do from other things that people say.

This is why it is so important to check that you understand the given instructions. Failure to do so could mean you carrying out the wrong activity or not doing something at all.

Remember If you are not sure of something, always ask for clarification.

Synoptic Practical Assignment

Bricklaying Level 1 – Construct straightforward brick and block walls

The synoptic practical assignment is a practical examination that takes place at the end of your course. It is an integral part of the Diploma and you must successfully complete it. Instructions for carrying it out are included on the practical assignment specification sheet. Marking of the test is carried out by your tutor by completing the marking sheet.

Information

For more information about bricks and other associated products, visit the websites of the better-known manufacturers and suppliers, or the Brick Development Association's website (www.brick.org.uk).

Quick Quiz

1. How many brick courses are there in 900 mm?

2. What does dry bonding entail?

3. What is the width of a one-brick wall?

4. List the three stages to risk assessment.

5. What are the two most common types of blocks used in the construction industry?

CHAPTER 11

Tools and Equipment

THIS CHAPTER RELATES TO UNIT CC1001K AND UNIT CC1001S.

The tools required by a bricklayer are few compared with many other trades, but care should be taken in their selection to obtain only the best quality, thus avoiding frequent renewals.

It is convenient to classify the tools required, together with a description of their uses, under the following headings:

- laying tools: brick trowel
- straightening tools: spirit levels, line and pins, corner blocks, straight edge, tingle plate and gauge staff
- cutting tools: club hammer, bolster, scutch, cold chisels, rule, square and bevel
- finishing tools: pointing trowel, jointer and hawk.

Laying Tools

Of the tools that a bricklayer uses, the brick trowel (Figure 11.1) is by far the most important as it is in almost constant use. In the process of laying bricks, the brick trowel is used to perform a series of operations, during which the trowel is seldom put down or changed from one hand to the other.

Straightening Tools

There are four principal straightening tools: spirit level, line and pins, straight edge and gauge staff.

Spirit Levels

These are metal straight edges specially fitted with glass tubes containing a spirit and a bubble of air. The type of level shown in Figure 11.2

Figure 11.1 Brick trowel.

Figure 11.2 Spirit levels.

is used for all general purposes, the length being about 1 m, but a useful asset in any kit of tools is the boat level, which is similar in all respects except length and shape, the former varying between 150 and 300 mm.

Line and Pins

The line used by a bricklayer is usually nylon and is wound round stainless steel pins, these pins being spear pointed at one end for easy insertion into the joints and flanged the other end to prevent the line slipping off the stem (Figure 11.3).

Corner Blocks

Corner blocks are used to attach the line to keep the brick or blockwork straight (Figure 11.4). They are made from wood, plastic or steel and fit on to the corners of the brick or blockwork with the line pulled tight to hold them in place. They are then raised to complete each course as it progresses.

Figure 11.3   Line and pins.

Figure 11.4   Corner blocks.

Straight Edge

Any length of timber that has parallel sides can be used for a straight edge, but for a better degree of accuracy a purpose-made straight edge should be employed (Figure 11.5). A straight edge is used in conjunction with a spirit level as a method of transferring a spot level

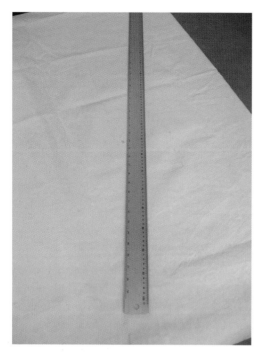

Figure 11.5 Straight edge.

Figure 11.6 Tingle plate.

from one point to another. When using the straight edge for levelling, it should be reversed end to end after each reading; this will do away with any inaccuracy in the straight edge or spirit level.

Tingle Plate

This is not an expensive tool but of great use. A tingle plate is used to support the centre of ranging lines on brick or block walls, removing the sag and thus giving better alignment (Figure 11.6).

Gauge Staff

This is a piece of timber of a length to suit its purpose, a convenient size for the bricklayer being 1 m × 50 mm × 25 mm; for workshop use it is usually marked off with saw cuts to the required gauge, that is one division on the gauge staff is equal to the height of a brick plus the joint (Figure 11.7).

The gauge staff used on a construction site is usually sufficiently long enough to reach from one floor to another. In this case it is known as a storey rod and, in addition to the brick courses, various important

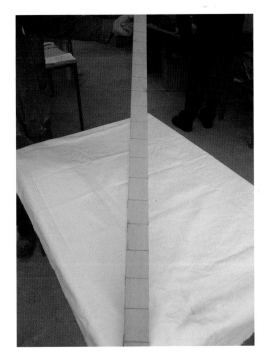

Figure 11.7 Gauge staff.

levels, such as sill level, head height, damp proof course and so on, are also marked on it.

Corner Profiles

Profiles are erected after three courses have been laid and allowed to go off over night. The purpose of profiles is to allow the construction of walls without the need to build up corners first. The time saved and greater accuracy have made corner profiles the accepted method of wall construction (Figure 11.8).

Method

Lines should be taken from corner profile to corner profile. The ideal configuration for a simple four-wall structure is to erect a profile on each corner and pass a line from profile to profile using the line holders and raise the line after each course. The line can be taken from pins or from corner blocks on the existing corner or doorway, but easier movement and better accuracy is achieved by mounting the line completely on the profiles.

Figure 11.8 Corner profiles.

Cutting and Marking Tools

The bricklayer makes use of a number of tools of this kind when plying their craft.

Club Hammer

The club hammer consists of a steel head fixed to a handle about 225 mm long. The head varies in weight between 1 and 2 kg, the hammer normally use by the bricklayer being 1.5 kg in weight. Figure 11.9 shows a club hammer, also known as a lump hammer in some parts of the United Kingdom.

Bolster

This is forged from steel and possesses a thin blade, the cutting edge of which has a slight convex curve (Figure 11.10). The blade, which may vary in width from 50 to 100 mm, is hardened for cutting, but the striking end should be left comparatively soft to avoid pieces of metal breaking and flying off when it is struck with the hammer.

Scutch or Comb Hammer

A scutch is used to trim bricks and blocks to the correct size or shape. There are several types of scutches, or comb hammers, on the market but the one illustrated in Figure 11.11 is the most common.

Brick Hammer

This type of hammer is used for the rough cutting of bricks. The chisel blade should always be kept sharp (Figure 11.12).

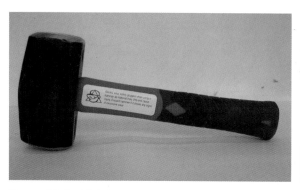

Figure 11.9 Club hammer.

Figure 11.10 Bolster.

Cold Chisels

These are made of varying lengths and thicknesses. They are forged from steel with the cutting edge and the striking end treated in a similar manner to those on a bolster (Figure 11.13).

Bricks reduced to various sizes and shapes are usually cut with the aid of a club hammer and bolster and trimmed to a fine finish, if required, with a scutch.

Cutting Gauge

Before cutting a brick, the position of the cut must be measured off and a square mark pencilled on with the aid of a try square. A specially prepared gauge may be used for marking off the lengths of closers, bats and three-quarter bricks (Figure 11.14).

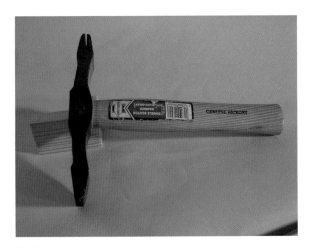

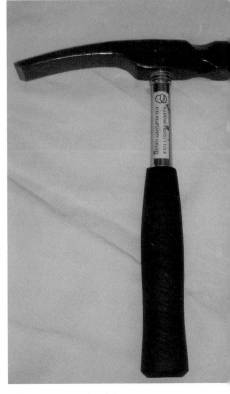

Figure 11.11   Scutch.

Figure 11.12   Brick hammer.

(a)

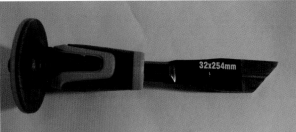

(b)

Figure 11.13   Cold chisels.

Figure 11.14   Cutting gauge.

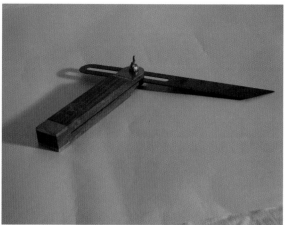

Figure 11.15   Bevel.

Bevel

A bevel is used for marking off an angle in the manner previously described for a square. The blade is pivoted at the top of the stock and fixed to enclose the required angle by fastening with a set screw (Figure 11.15).

General Finishing Tools

For putting the final touches to their work, bricklayers need a number of special tools.

Pointing Trowels

These are similar in appearance to the ordinary brick trowel but the blades are proportionally smaller and usually made of a lighter gauge steel, the length of blade varying between 50 and 150 mm. It is usual practice for the bricklayer to carry two pointing trowels, one about 50 mm long, known as a cross-joint trowel, the other about 150 mm long, and known as a bed joint trowel (Figure 11.16).

Hawk

The bricklayer's hawk is usually made in two parts for convenience in carrying, a top portion which is about 150 mm square and a handle 100 mm long. The handle is attached to the top by means of a dove-tailed slip, which slides easily into a prepared groove, thus forming a satisfactory and rigid connection (Figure 11.17).

The hawk is used to hold mortar for pointing and forms a convenient place from which material may be transferred to the trowel.

Figure 11.16 Pointing trowel.

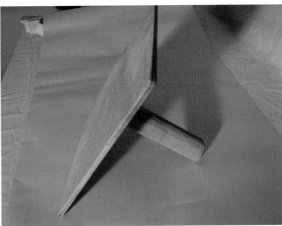

Figure 11.17 Hawk.

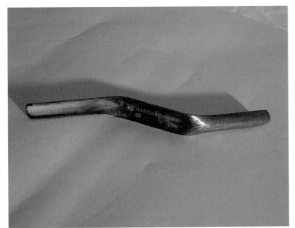

Figure 11.18 Jointer.

Figure 11.19 Tape measure.

Jointer

Jointers are usually made from steel. The tool is used to form the joint finish during building the wall. This serves two purposes: it helps to make the joint more waterproof and the appearance of the wall looks more pleasing.

The jointer illustrated in Figure 11.18 is used to form a half round joint, often referred to as a bucket handle finish.

Tape Measure

For maximum accuracy, always use a steel tape measure (Figure 11.19). The majority of steel tapes are graduated into metres, centimetres and millimetres and can be read to the nearest millimetre. Tapes are usually available in lengths of 3, 5 and 10 m.

Equipment for Working with Mortar

Mortar Tubs

Mortar tubs are containers used for hand-mixing mortars or for storing machine-mixed mortar, the capacity of mortar tubs range from about 1 to three cubic metres. They are usually constructed of metal or plastic (Figure 11.20).

Mortar Boards

After the mortar has been mixed, it must be placed where the brick-layer can pick it up for spreading. Mortar boards serve this purpose (Figure 11.21). They can be made from steel, wood or synthetic material.

Mortar Mixers

When a continual supply of mortar is required, it is sometimes necessary to machine-mix for the sake of economy. The loading capacities of mixers vary from 50 kg to upwards of 150 kg and can be powered by electric motor or petrol/diesel engine. Job conditions will determine the capacity of mortar required. The availability of electricity or fuel will determine your choice of mixer (Figure 11.22).

Figure 11.20   Mortar tubs.

Figure 11.21   Mortar board.

Figure 11.22 Mortar mixer.

Mobile Hoists

On large, economically operated construction sites, mobile hoists (Figure 11.23) are sometimes employed to place material on scaffolds or floors above ground level. Materials can then be quickly positioned for use by the bricklaying team.

Figure 11.23 Mobile hoist.

Power Tools

Angle Grinder

Angle grinders are cutting tools that run by electricity using 110V and 230V supplies, or are battery powered. They cut using an abrasive or diamond-type disc. They range from 100mm to 230mm disc size.

They are used by bricklayers for cutting bricks, blocks, concrete and stone to size or for cutting existing material for alteration. Owing to the cutting speed, large amounts of dust are released from the material, so goggles and a mask should always be worn in addition to the usual personal protective equipment (PPE). Ear defenders should also be worn. All leads should be checked before use and after use for cuts or splits, and with a 110V supply a transformer must be used.

Block Cutters

These are used for cutting bricks and blocks and have a sharp cutting edge and pressured compression to break the material to the correct size.

Masonry Saw and Blades

A masonry saw is designed for cutting masonry material, stone, concrete block, brick and tiles and so on. This piece of equipment is usually powered by electric motor. It can be fitted with a water pump, which is used for wet cutting. Other types of blades for dry cutting are manufactured with abrasive materials on the cutting surface or edge.

Maintenance of Tools

The tools required by a bricklayer are few compared with many other trades, but care should be taken in their selection to obtain only the best quality, thus avoiding frequent renewals and the necessity of acquiring the 'feel' of new tools at frequent intervals.

Tools of any description deserve careful treatment; the reward to the owner is a prolongation of their life and greater efficiency and ease when working. Owing to the nature of the work, bricklayers' tools can easily spoil through rusting or, in the case of timber tools, through constant changes of climatic conditions. It should, therefore, become a habit thoroughly to clean and dry all tools after each day's work.

Information

For more information about tools and equipment, visit the websites of the better-known manufacturers and suppliers.

1. List five cutting tools.

2. Describe the method used for erecting corner profiles.

3. What is a tingle plate used for?

4. Of the tools that a bricklayer uses, what is the most important?

5. What is a bevel used for?

CHAPTER 12

Pointing and Jointing

THIS CHAPTER RELATES TO UNIT CC1003K AND UNIT CC1003S.

Introduction

Pointing is the term applied to the process of raking out the jointing mortar before it has set hard and filling the joints with a mortar of a different character, typically to a depth of 15 mm, to create a specific finish. Pointing is not to be confused with jointing, which means forming the finished surface profile of a mortar joint by tooling as the work proceeds, without pointing. Repointing is the raking out of old mortar and replacing it with new.

Pointing

The original object of pointing was to protect the bedding material of the brick and consequently the bricks from the effects of the weather. Recently, it has been used as a means of varying the appearance of the brickwork.

There are several types of pointing and these may be carried out in a variety of colours by means of coloured cements and dyes and the use of different sands, the combination making possible an almost unlimited range of finishes.

Sequence for Pointing

Pointing is one of the most difficult skills to become proficient in and it requires a great deal of practice to achieve a satisfactory finish.

There is a sequence to pointing that should always be followed.

Stage One

Always start at the highest part of the wall.

Stage Two

Taking about one square metre ($1\,m^2$) at a time, remove any mortar stains or loose mortar from the joints and face of the brick; care must be taken not to damage the face of soft bricks.

Stage Three

Wet the wall, starting at the top. The amount of water used will depend upon the absorption rate of the bricks. Allow water to be fully absorbed from the face of the wall before beginning the pointing.

Stage Four

Using a hand hawk and small pointing trowel, apply mortar to the cross joints first. Finish each joint by compressing and ironing the mortar with the trowel until it is slightly indented on the left hand side of the joint. Complete about $1\,m^2$ at a time.

Stage Five

In the same manner, fill the bed joints and compress and iron in with the top of the joint slightly cut back behind the arris of the brick above.

Stage Six

Once the joint has set, it should be lightly brushed to remove any surplus mortar.

Forming Joint Finishes in New Masonry Work

Words and Meanings

Tooling – The process of compacting, smoothing and sealing horizontal and vertical mortar joints.

In order to prevent water from penetrating a block, brick or stone wall through the mortar joints, it is necessary to seal the joints. Tooling the joints will seal them.

Tooling mortar joints results in shaped joints:

- concave shape (Figure 12.1)
- recessed shape (Figure 12.2)
- flat shape (Figure 12.3)

Unless otherwise specified, all mortar joints are tooled to concave shapes. Mortar joints should be tooled after the mortar has stiffened but before it has hardened.

The length of time it takes for the mortar to harden will depend upon:

- weather conditions (mortar hardens quicker in warm weather than in cold weather);
- mortar hardens slowly when using glazed or well-burnt bricks;
- mortar hardens quickly when using porous bricks.

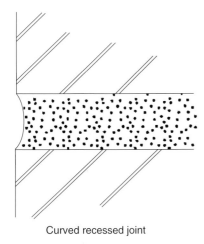

Curved recessed joint

Figure 12.1   Concave joint.

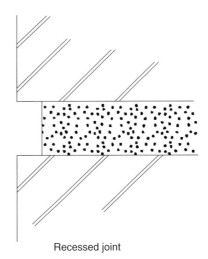

Recessed joint

Figure 12.2   Recessed joint.

The ability to judge or determine the condition of mortar joints will come from practical experience, such as learning to check the readiness of the mortar using your fingertips. Ask your tutor to show you this method.

The jointer is used to seal the horizontal joints between bricks and blocks. The usual types of jointer are either:

- half round or
- round.

Tooling of the vertical joints is carried out using a small S- shaped jointer (Figure 12.4).

Practise the following procedure and technique.

- First tool the vertical joints (Figure 12.5). Place the S-shaped jointer at the top of the vertical joint and move it downwards in a straight line.
- Fill voids with fresh mortar and retool (Figure 12.6). Repeat until the joint is smooth and free of voids.
- Next, seal the horizontal joints by moving the jointer back and forth along the top edge of the bricks (Figure 12.7). Again, fill voids and seal.
- Lightly retool the vertical joints. Use your trowel to remove all loose material from the face and base of the wall.

Ask your tutor to show you samples of jointing tools used locally.

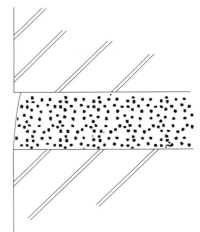

Weather struck joint

Figure 12.3   Flat joint.

Figure 12.4   Jointer.

Figure 12.5   Tooling vertical joints.

Figure 12.6   Filling joints with mortar.

Figure 12.7   Tooling horizontal joints.

Mixes for Pointing

The selection of the aggregate and matrix will depend upon the type of pointing and the finish required, but the mortar used should not be excessively strong in comparison with the brick, as there is a tendency for the brick to wear away and leave the joints projecting.

Mixing

To form a mortar that may be easily used, all the materials should be passed through a fine sieve and be thoroughly mixed in a dry state before any water is added. Since, for pointing, the mortar is taken up on the back of the trowel, it is desirable to have it much stiffer than ordinary mortar; therefore, a smaller amount of water is required and the mortar is brought to its correct consistency by beating the materials well together with the back of a shovel. It must also be remembered that an excess of water tends to render the joints porous.

Mix Ratios

The matrix may be cement, lime or a combination of the two, and it is mixed with the aggregate in varying proportions according to the finish required. The following proportions are usually adopted:

Weather Pointing

- 1 part cement, 1 part sand for engineering bricks
- 1 part cement, 2 parts sand for general building work.

Flush Pointing

- 1 part cement, 2 parts sand or
- 1 part cement, 1 part lime, 3 parts sand.

Coloured Joints

To obtain a coloured joint, a proprietary brand of pointing material is used that only requires mixing with water before using. Because of the wide range of finishes that can now be obtained by varying the proportions and materials, it is general practice to prepare small panels of brickwork pointed in a number of different shades of the colour required, so that the architect may make a selection.

Water

Water used in the preparation of mortar should be clean and free from impurities, which may impair the necessary characteristics of good mortar. Impurities such as salt and oil are harmful since they reduce the adhesive property of the mortar.

Joint Profiles

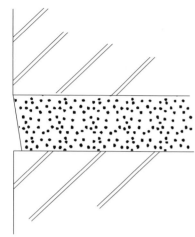

Struck joint

Figure 12.8 Struck joint.

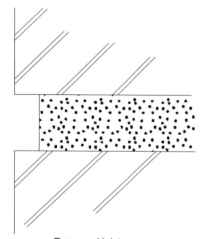

Recessed joint

Figure 12.9 Squared recessed.

The long-term performance of brickwork is highly dependent on the correct mortar joint profile for the efficient shedding of rainwater. Brickwork that remains saturated is more susceptible to frost damage. The choice of joint profile should therefore be based on performance criteria, as well as aesthetic considerations.

Weather struck and bucket-handle joints assist rainwater run-off, yet give some definition and so are often considered an optimum solution. Recessed joints create shading that will emphasise the bricks, whilst flush joints have the reverse affect. Special joints may require special jointing tools.

Joint finishes carried out during the process of building vary considerably, the principal ones being described below.

Struck

This type of joint finish is largely confined to (a) internal brickwork and (b) overhand building and finishing, and is formed by pressing in the joint with the back of the trowel point, the cross-joints being ironed in first. The ease with which this may be carried out will depend on the fullness of the joints as the work is built, thereby reducing to a minimum any necessity for filling the joints afterwards. See Figure 12.8.

A struck joint is not suitable for external use, as the base of the joint will hold water.

Squared Recessed

Very attractive profile – not generally recommended for free-standing walls or any exposed situations. The depth of recess should be kept to the minimum necessary to achieve the desired appearance, but should not be any greater than 6mm. It is achieved by raking out the mortar joint with a chariot recessing tool (Figure 12.9).

Weather Struck

This type of joint produces a contrasting effect of light and shade on the brickwork. Such joints, when correctly formed, have excellent strength and weather resistance and are suitable for all grades of exposure (Figure 12.10).

Flush

This joint gives maximum bearing area for the brick and is often favoured when coarse-textured bricks are used. With some brick types, the finish may appear a little irregular. This is suitable for moderate and sheltered exposures (Figure 12.11).

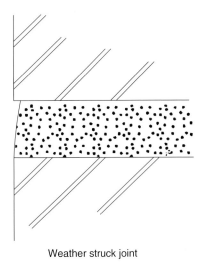

Weather struck joint

Figure 12.10 Weather struck.

Flush joint

Figure 12.11 Flush joint.

Curved Recessed

This type of joint gives an improved appearance over a flush joint with little reduction in strength. Owing to the compression of the joint and the superior bond, it has good weather resistance and is suitable for all grades of exposure (see Figure 12.1: a concave joint and a curved recessed are different terms for the same joint).

Information

For more information about pointing and jointing, visit your local or college library and borrow textbooks on the subject.

Quick Quiz

1. What is the difference between pointing and jointing?

2. Define the term 'tooling'.

3. Describe the method for mixing mortar for pointing.

4. Sketch four joint profiles.

5. How are coloured mortar joints obtained?

CHAPTER 13

Bonding

THIS CHAPTER RELATES TO UNIT CC1003K AND UNIT CC1003S.

Bonding Arrangements

While it is generally recognised that bricklaying is a manipulative art calling for skill in the handling of tools, its practice also requires a complete understanding of the correct arrangement of bricks to form a wall. This arranging of the bricks is known as bonding, the bricks being so placed that they interlock one with the other, and care being taken to see that, as far as possible, no vertical joint between them is immediately over a vertical joint in the course below.

Bonding of Walls

Figure 13.1 shows a beam resting on a wall built without bond, that is a wall in which the connection between the various brick units is dependent upon the adhesive properties of the mortar only. With a load from a beam such as this imposed on any part of the wall, the tendency is for that part to subside, fracturing the mortar joints and leaving the remainder of the wall undisturbed. To avoid this strain and probable fracture being set up in continuous vertical mortar joints, it is necessary to intercept or bridge them with bricks, causing the bricks in the respective courses to overlap and thereby creating a bonded wall.

Figure 13.2 shows the same beam bearing on a wall that is properly bonded and consequently has no continuous vertical joints, the result being that instead of the load pressing directly down on one column of bricks it is spread out along the lines of the joints, thus distributing it over many more bricks and so over a much greater area.

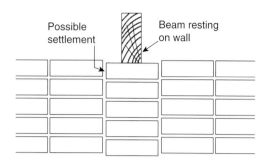

Figure 13.1 Wall built without bond.

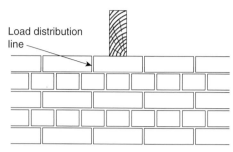

Figure 13.2 Wall built of bonded brickwork.

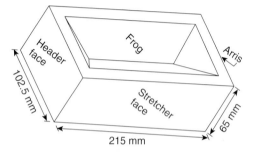

Figure 13.3 Stretcher and header faces.

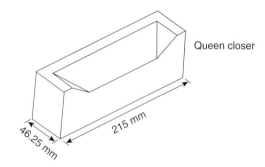

Figure 13.4 Queen closers.

Types of Face Bonds

- **Stretcher:** A brick laid flat with its long face parallel to the wall (Figure 13.3).
- **Header:** A brick laid flat with its header face showing in the wall (Figure 13.3).
- **Queen closer:** Closers are formed from cut bricks, the surface exposed on the face of the wall measuring 46.25 × 65 mm. There are several types of closer but the two in general use are the ordinary closer and the queen closer, as illustrated in Figure 13.4.
- **Bat:** This is obtained by cutting a brick through the centre line and across the width (Figure 13.5).
- **Three-quarter:** This is also a cut brick (Figure 13.6).

Types of Bond

There are several types of bond; some are employed to obtain maximum strength, while others are used for their appearance or economy. The arrangement of bricks to suit various lengths of wall requires very

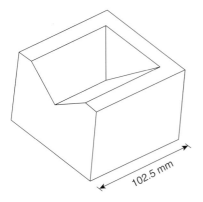

Figure 13.5 Brick bat.

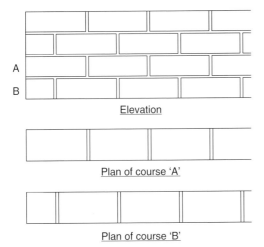

158.75 mm

Figure 13.6 Three-quarter brick.

Elevation

Plan of course 'A'

Plan of course 'B'

Figure 13.7 Stretcher bond wall.

Figure 13.8 Section of a one-brick wall if built of two half-brick walls.

careful consideration, for while the general sequence of bond must always be followed it will be seen that variations are certain to occur according to the length and breadth of the wall required.

Stretcher Bond

The simplest form of bonding is that for a wall 102.5mm thick (known as a half-brick wall), where all the bricks are laid down as stretchers, each lapping over the one below by half its length (Figure 13.7), which shows an elevation and plans of the two alternate courses on the wall.

This form of bonding would, however, only be suitable for the 102.5mm wall, since if it were required to increase the thickness of the wall to 215mm there would be no connection between the two 102.5mm walls built face to face, as may be seen from the section shown in Figure 13.8.

Figure 13.9 Section of one-brick wall built in normal manner.

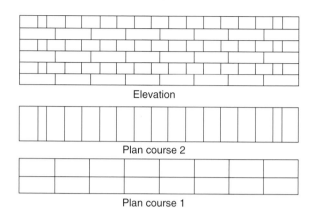

Figure 13.10 English bond.

It is therefore necessary to introduce ties across the two walls in the form of headers, and in this way a bonded wall 215 mm thick is formed, built of alternate courses of headers and stretchers as illustrated in Figure 13.9. It is clear that as it is necessary to bond a wall along its length, so it is also necessary to bond it across its thickness in order that the entire mass may be tied together and the distribution of load take place in both directions, that is along its length and across its thickness.

Bonding, therefore, is the systematic arrangement of bricks in a wall whereby each brick overlaps other bricks in the courses below in both directions, so distributing, over a wider area, a load imposed on any part of the wall. While the principle of bonding remains constant, the arrangement of bricks in a wall may be varied considerably, hence we have different names prefixing the term 'bond', each name denoting a recognised form of face bonding which is distinguished from the others by its elevation. Of the many types of bond used, the most common is half-bond, or stretcher bond.

English Bond

This consists of alternate courses of headers and stretchers with the centre of the stretcher in line with the centre of the header in the courses above and below. The header course is commenced with a quoin header followed by a queen closer and continued with successive headers. The stretcher course is formed, on its outer face, of stretchers having a minimum of one-quarter their length over the headers. See Figure 13.10.

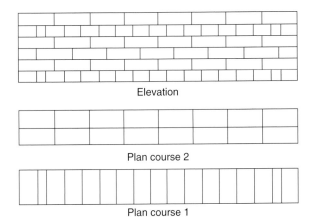

Figure 13.11   English garden wall bond.

English Garden Wall Bond

English garden wall bond consists of a course of headers, with the necessary queen closer next to the quoin header. There are then three, or sometimes five, courses of stretchers in series running the full height of the wall. See Figure 13.11.

Flemish Bond

This consists of alternate headers and stretchers in the same course with the centre of the stretcher in line with the centre of the headers in the courses above and below. In order to obtain the lap, which is one-quarter the length of the bricks, a queen closer is introduced next to the quoin header in alternate courses followed by a stretcher. The succeeding course commences with a stretcher followed by a header, which is placed centrally on the stretcher below. See Figure 13.12.

Flemish Garden Wall Bond

Flemish garden wall bond consists of alternate courses composed of one header to three, sometimes five, stretchers in series throughout the length of the courses. The lap is obtained by either introducing the necessary queen closer next to the quoin header or by use of a three-quarter bat. Headers other than quoin headers are placed centrally on the centre stretcher of the series. See Figure 13.13.

Advantages of English and Flemish Bonds

It is generally considered that English bond is stronger than Flemish bond, because there are no vertical joints within the mass which stand over each other in successive courses. The one advantage of Flemish bond is in its rather more decorative appearance.

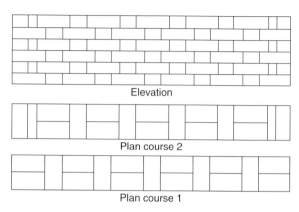

Figure 13.12   Flemish bond.

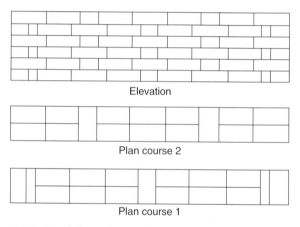

Figure 13.13   Flemish garden wall bond.

Return Corners

All changes in the direction of walls are termed return angles, and these may enclose an obtuse or acute angle or, more commonly, a right angle. When considering the bonding to a right-angle return, it is simpler to think of two separate straight walls lapping one over the other on alternate courses and forming a right angle; this arrangement is shown in Figure 13.14. It follows, therefore, that the rules applicable to straight lengths of wall will also apply to return angles with only minor adjustments to be made.

Developing the outline shown in Figure 13.14 and arranging the bricks in stages as for straight lengths of wall, first set out those bricks numbered 1 in Figure 13.15, noting that where the wall changes direction the bond in the same course also changes.

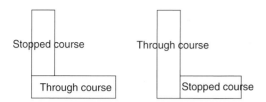

Figure 13.14 The lapping of walls.

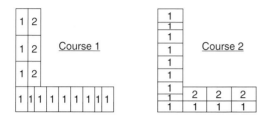

Figure 13.15 Courses 1 and 2 in English bond.

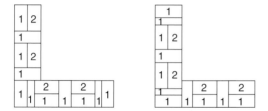

Figure 13.16 Courses 1 and 2 in Flemish bond.

Following on to the next stage and filling in those bricks numbered 2 in Figure 13.15, we arrive at the arrangement shown, thus completing the bonding for this example. Figure 13.16 illustrates an example in Flemish bond treated in a similar manner.

The bonding to return angles presents very little extra difficulty than that of straight lengths of wall, but the other examples (shown in Figures 13.17 and 13.18) should be studied to appreciate this fully.

With bonding to the internal angle, there are, however, certain points requiring particular consideration when the brickwork is facework. In the case of English bond for walls containing an odd half-brick in their width, it is usual practice to allow the tie stretcher brick at the back of the course to enter the wall by a quarter leaving three-quarters projecting outwards along the course and thus avoiding a mass of joints near to the angle.

With Flemish bond, the sequence of the bond should be followed as closely as possible but, at all times, the bricks should be so arranged that the centre of the header in one course is over the centre of the

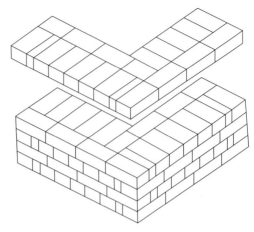

Figure 13.17   Return angle: English bond.

Figure 13.18   Return angle: Flemish bond.

stretcher below. Bonding one course without any reference to the other may easily result in this essential point being overlooked.

Setting Out a One-brick (English Bond) Wall

Set out in English bond a 215 mm wall, usually referred to as a one-brick wall, 890 mm long, the steps being illustrated in Figure 13.19.

Before setting out the bond, it is necessary to draw two outlines of the plan of the wall which, for the example in English bond, will be two rectangles 890 mm × 215 mm, a plan being required for both the header and stretcher course.

The next step is to bond the face of the wall starting from each end, carefully noting that, irrespective of the length of the wall to be bonded,

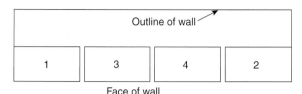

Face of wall

Stage 1   Set out stretchers on face in order shown.

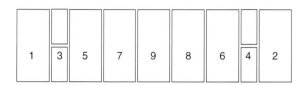

Stage 2   Set out heading course in order shown.

Stage 3   Complete bricks to back of stretching course

Figure 13.19   Setting out an English bond wall.

the bricks on the corners must be placed in the same direction, that is if a stretcher is placed on one end then a stretcher must be placed on the opposite end. The same applies to the header course.

Working towards the centre and numbering the bricks in the order in which they are placed in position, it is found that the face of one course consists of four stretchers.

On the second course or the header course, a closer is introduced next to the quoin header to obtain the necessary quarter lap, the space between the closers being then filled with headers. The face bond is now clear for the stretcher and header courses of a wall 890 mm long, built in English bond.

To proceed with the bonding to the back of the wall, a 215 mm wall being two half-bricks thick will obviously have stretchers on both sides in the same course; hence the back of the stretcher course will be filled in with stretchers in the order and manner illustrated. Because the width of a 215 mm wall is governed by the length of a header, the bonding to the header course is complete.

Flemish Bond Wall

Now apply the same method to a similar example in Flemish bond – that is to a 215mm wall, 890mm long. The stages are illustrated in Figure 13.20.

Outline the two rectangles in which the bricks are to be arranged, and set out the bond along the face of the wall. Place the two quoin stretchers in position, and continue with headers and stretchers alternately from each end towards the centre, thus arriving at the arrangement shown at Stage 1. To bond the next course, start with headers at either end, followed by closers, and fill in towards the centre with alternate stretchers and headers in the order and manner shown in Stage 2.

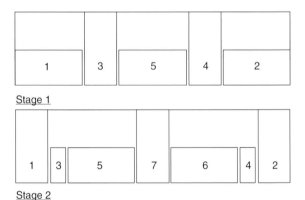

Stage 1

Stage 2

Stages 1 & 2   Set out bricks to face of each course in order shown.

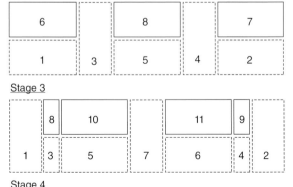

Stage 3

Stage 4

Stages 3 & 4   Complete bricks to back of wall as shown

Figure 13.20   Setting out a Flemish bond wall.

The face bond is now completed for the two courses in Flemish bond, which, like English bond, will remain the same for walls of any thickness, providing the length is constant.

Arrange the bricks to the back of the stretcher course and consider the corner bricks first; to comply with the rules of bonding, these must be stretchers, the remaining space being obviously filled in with a stretcher, as shown at Stage 3.

With the rules of bonding again as a guide to the position of the transverse joints, the correct arrangement for the remainder of the bricks in the header course is decided and shown in Stage 4.

Bonds for Junctions

This is the term given to connections between walls which form the shape of a letter T, or a cross on plan, although it is not essential that the angles enclosed be right angles. The bonding is treated in a similar manner to that for attached piers, that is the bond for the main length of wall is first set out and the adjoining wall tied in at the most convenient course which, in English bond, is usually found to be in the stretcher courses.

Figures 13.21 and 13.22 show examples of English and Flemish bond for walls of different thicknesses. Similar principles are applied

Figure 13.21 Junction in English bond.

Figure 13.22 Junction in Flemish bond.

Figure 13.23 Cross junctions.

Note

Blockwork bonding is covered fully in Chapter 8.

to the bonding for walls that cross, some examples being illustrated in Figure 13.23.

Information

For more information about bonding, visit your local or college library and borrow textbooks on the subject.

Quick Quiz

1. Define the term 'bonding'.

2. Draw an English and Flemish bond wall.

3. List five different face bonds.

4. What is a queen closer?

5. List the stages in setting out an English bond wall.

CHAPTER 14

Materials

THIS CHAPTER RELATES TO UNITS 1014K, 1014S, 1015K, 1015S, 1016K AND 1016S.

Introduction

Ordinary bricks are unit blocks manufactured from clay, concrete or calcium silicate. The characteristics and colours of bricks vary with the method of manufacture, the composition of the material used and the natural position from which the material is obtained.

We can list the difference between types of bricks according to:

- method of manufacture
- classification by use.

Method of Manufacture

Bricks may be broadly subdivided under the following headings based upon the process of making and firing.

Handmade Bricks

As the name suggests, these bricks are entirely handmade. The mould in which the bricks are made consists of a bottomless box that fits over a board having the frog of the brick formed in the reverse upon it. Clay is thrown into the mould, which has been previously wetted or sanded to prevent the clay adhering to the sides. The surplus clay is cut off with a piece of wood known as a striker, or with a length of wire stretched over a wooden frame. The box is then removed leaving the brick ready for drying.

Wire-cut Bricks

These bricks are partially machine-made, the clay in a suitably plastic state being forced by revolving blades through a rectangular opening, which is the length by the width of a brick plus a shrinkage allowance, in one continuous length on to a steel table. A frame containing several wires spaced the thickness of a brick apart is brought down across the clay, cutting it into a number of pieces, each piece being the size of a brick before baking. The brick has no frog and the wire marks can be seen on both beds of the brick.

Pressed Bricks

If it is desired to give bricks that have been wire cut a sharp arris and a frog, the process of cutting off by wire is followed by machine pressing. The bricks are conveyed to a metal mould, the sides and top of which are simultaneously compressed by mechanical means. This produces a good sound brick that is regular in shape and size.

Each of these processes produces in the finished brick certain characteristics, which largely govern the use to which the brick is put: general construction, decorative work, engineering work and so on.

Clay Bricks

Clay bricks are usually pressed, cut or moulded and then fired in a kiln at a very high temperature. Their density, strength, colour and surface texture will depend on the variety of clay used and the temperature of the kiln during firing.

Concrete Bricks

These types of bricks are made in a similar way to sand lime and flint lime bricks, except that sand and cement are used instead of sand and lime.

Sand Lime and Flint Lime Bricks

Sand lime and flint lime bricks, or calcium silicate bricks, as they are more properly called, are made from a mixture of sand, crushed flint, pebbles or rock, or a combination of such materials, with hydrated lime. They are moulded under pressure and hardened by exposure to steam at very high pressure. On cooling, they are ready for immediate use.

Classification by Use

Bricks may be classified according to their use as follows, bearing in mind that it is sometimes possible for a brick to come under more than one heading.

Facing Bricks

Facing bricks are intended to provide an attractive appearance. They are available in a wide range of types, colours and textures. Some may not be suitable in positions of extreme exposure. Some facing bricks also have engineering properties.

The term 'facing' covers a very wide range of bricks since it includes all those used for exterior and interior walls that are to be left as finished work.

Common Bricks

Common bricks are suitable for general building work and are not chosen for their appearance.

These are bricks for ordinary work that is not exposed to view, for example walls that are to be plastered or are built underground. Nearly all brick makers produce common bricks, which are often manufactured from the same clay as that used for better-class products but which lack the finer preparation and finish.

Engineering Bricks

Engineering bricks are hard-burnt bricks that are very dense. They have a guaranteed compressive strength and minimum water absorption. They are not chosen for appearance and there is no requirement for colour.

These are bricks suitable for ground works, manholes and sewers, retaining walls or as ground level damp proof courses for free-standing walls, or in situations where high strength and low water absorption are the most important factors.

Special Bricks

A wide variety of bricks are available in special shapes or sizes, to blend or contrast with most facing bricks.

Brick Dimensions

The dimensions and shapes of bricks vary according to the requirements of the purpose that the bricks are intended to serve. Where bricks varying in size are used together, the task of the bricklayer is made difficult on account of the additional care required to produce work of good appearance. To overcome this difficulty, bricks should have standard dimensions, but even the most up-to-date methods of manufacture inevitably produce slight variations in size.

The current British Standard brick size is 215 mm long, 102.5 mm wide and 65 mm thick.

Table 14.1   Quantities of Bricks and Mortar

Wall thickness (mm)	No. of bricks required per square metre of wall	Volume of mortar per square metre of wall (m³)		Volume of mortar per 1000 bricks used (m³)	
		frog up	frog down	frog up	frog down
102.5	60	0.031	0.021	0.52	0.35
215	119	0.072	0.052	0.61	0.44

The mix proportions given contain the minimum recommended cement content for durability, if for any reasons mortar with greater cement contents are required, then stronger mortars may be satisfactory with the London range of bricks.

The BS 5628: Part 3 table 15 mortar designation is given for each recommended mortar mix to assist those involved in structural design calculations to BS 5628: Part 1 and 2.

For work below finished ground level, the mortar may have to be varied, depending on the level of sulphates in the soil or ground water. For details, see table 3. The same mortar should be used for all work up to ground level DPC, or at least two courses above finished ground level.

Free-standing walls, parapet walls and chimneys must be finished with an overhanging coping.

Retaining walls must have a water-proof backing and an overhanging coping.

Approximately 60 bricks are required per square metre of half-brick walling and 120 for full-brick, including 10 mm mortar joints. Table 14.1 may be useful to calculate the numbers required and the mortar needed – but always remember to subtract an allowance for window and door openings.

Definition of Sizes

- Coordinating size: The size of a coordinating space allocated to a brick, including allowances for joints and tolerances.
- Work size: The size of a brick specified for its manufacture, to which its actual size should conform to specified permissible deviations.
- See Figure 14.1.

Dimensional Variation

During the brick manufacturing process, there is sometimes considerable change in the size of the newly formed brick, as it dries out. For this reason the exact sizes of bricks are difficult to control.

British Standard BS EN 771 specifies brick size and tolerances, based on the measurement of 24 bricks – see the chart in Figure 14.2.

Sizes

	Length mm	Width mm	Height mm
Coordinating size	225	112.5	75
Work size	215	102.5	65

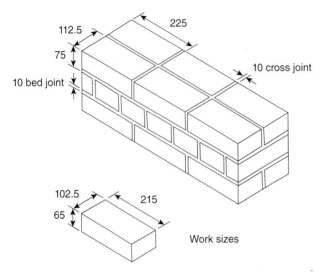

Co-ordinating sizes

225
112.5
75
10 cross joint
10 bed joint

102.5
215
65

Work sizes

Figure 14.1 Coordinating and work sizes.

Size

Work Size (mm)	Tolerance – Measurement of 24 bricks		
		Max (mm)	Min (mm)
Length: 215	Length	5235	5085
Width: 102.5	Width	2505	2415
Height: 65	Height	1605	1515
Compressive Strengths: ➤ 20 N/mm^2 frog up ➤ 7 N/mm^2 frog down (these strengths comply with the special category of manufacturing control in British Standard 5628: Part 1 Clause 27.2.1.2)			
Durability Designation: MN – Moderately frost resistant, normal soluble salts content			
Water Absorption > 12%, typically 22% Initial rate of suction – typically 2kg/m^2 minute			

Figure 14.2 Size chart.

Durability and Position in the Build

Bricks are classified according to frost resistance and soluble salt content. They are given different designations for durability. Durability is defined as 'capable of lasting – resisting wear'. External claddings are expected to meet these criteria and brickwork is no exception.

Durability Designation

All clay bricks have a durability designation rating and it is important to know which bricks to use and where.

Bricks produced in the United Kingdom fall into three categories of frost resistance (see Table 14.2):

- F – frost-resistant
- M – moderately frost-resistant
- O – not frost-resistant.

F-rated bricks can be used in all normal building situations and degrees of exposure.

M-rated bricks are also durable except where they remain saturated and are subjected to repeated freezing and thawing. Generally, they can be used between the damp proof course (DPC) and eaves, although caution should be exercised on sites in elevated/exposed locations.

Do not use M-rated bricks:

- below ground-level DPC
- for sills
- for copings and cappings
- beneath cappings
- for projecting details (e.g. plinths)
- in exposed site locations

O-rated bricks should not be used externally.

Note

Bricks do not look any different, so ask your supplier for the rating. If bricks are not rated, they should be assumed to be rated O: not frost-resistant.

Table 14.2 Durability chart

Designation	Frost-Resistant	Soluble Salts Content
FL	frost-resistant (F)	Low (L)
FN	frost-resistant (F)	Normal (N)
ML	moderately frost-resistant (M)	Low (L)
MN	moderately frost-resistant (M)	Normal (N)
OL	not frost-resistant (O)	Low (L)
ON	not frost-resistant (O)	Normal (N)

Types of Blocks

There is a wide variety in the size and type of block used within the construction industry. Some of these are described below.

Lightweight Insulation Blocks

Aircrete is a lightweight, load-bearing and thermally insulating building material most commonly available in block format, but also as reinforced units. The blocks are light in weight, easy to work and have excellent load-bearing capabilities.

Dense Concrete Blocks

The aggregate concrete block is by far the most commonly used building block type in the construction industry, representing almost 70% of new construction.

Concrete blocks are strong, durable, ideal for all wet finishes and for dry lining and are easy to lay and position on mortar. Concrete blocks are available in various strengths, weights, sizes and surface textures.

Lime

Lime is a fine-powdered material, with no appreciable setting and hardening properties, used to improve the workability and water retention of cement-based mortars.

Cement

A fine-powdered material which, when mixed with water, sets and binds together to form a hard solid material, cement is used as a component of mortar and concrete. In the United Kingdom, the most commonly used cements in mortars are Portland cements and masonry cements.

Types of Cement

Ordinary Portland Cement

Ordinary Portland cement is the most widely used cement for application where chemical or thermal attack is not to be encountered.

Masonry Cement

Masonry cement has air-entraining and plasticising properties and produces a mortar having a high workability.

Figure 14.3 Fine aggregate.

Sulphate-resisting Cement

Portland cement is liable to sulphate attack in damp conditions; therefore, the composition of sulphate-resisting cement is adjusted so that it withstands sulphate action.

Rapid-hardening Portland Cement

Where greater speed is required in the final setting time of concrete, rapid-hardening cement may be used.

High-alumina Cement

This type of cement hardens much more quickly than those previously mentioned do. The initial set is completed in two or three hours and the first hardening in one day; also, the ultimate strength is greater.

Fine Aggregate

Fine aggregate or sand used in mortar for brickwork should be free from all earthy and organic matter and not too fine. While it should generally be fairly sharp or gritty, an addition of soft sand will make the mortar easier to manipulate (Figure 14.3).

Mortar

The material in which bricks are usually bedded to form a wall is known as mortar. Mortar is a mixture of sand and cement or lime, or all three, which hardens as it dries. Mortar is an essential ingredient of brickwork and is subject to the same exposure as the brick; unfortunately, it is not often given the same degree of consideration as the brick. Good workmanship and a thoroughly mixed mortar will contribute to durable construction.

Lime Mortar

This usually consists of one part lime to three parts of sand by volume, but may vary in its proportions according to the type for which it is intended.

Cement Mortar

This is a mixture of Portland cement and sand in proportions that will vary with the nature of the walling to be built. For all general purposes, this is usually one part cement to four parts sand by volume.

Pre-mixed Mortars

These mortars are produced in a quality-controlled environment to ensure accurate mix proportions. There are two main types available: ready-to-use and lime/sand mortars.

Ready-to-use Mortars

These mortars are produced in a factory and delivered to site ready to use. They may be:

- wet ready to use, which requires no further mixing and is stored in tubs on site;
- dry ready to use delivered in silos or bags, which requires only the addition of mixing water.

Lime/sand Mortars

These are pre-batched materials that are delivered to site, with cement and water being added prior to use.

All production methods of factory-produced mortars can offer both coloured and natural shades.

Concrete

Concrete is composed of a matrix and aggregates of varied size, mixed with the correct amount of water to a plastic condition. While still plastic it may be moulded or cast to any desired shape, but after setting it becomes extremely rigid.

Coarse Aggregates

Aggregates for concrete form the body of the mixture, as in mortar. The aggregate for ordinary concrete may vary from sand to broken (or natural) material of approximately 35 mm diameter. The sand is often called fine aggregate and the rest coarse aggregate (Figure 14.4).

Figure 14.4   Coarse aggregate.

Water

All water used in the mixing of concrete should be clean and free from organic or mineral impurities.

Checks on Materials

Bricks and Blocks

On delivery, check all documents and paperwork against original orders and specifications to ensure that:

- the delivery is of the correct brick and block type, size and quantity;
- the bricks and blocks are within the specified limits of size. If in doubt, inform the supplier immediately;
- any variation in colour is within acceptable limits, particularly if the delivery contains brick or block specials. If in doubt, inform the supplier immediately;
- any minor blemishes, such as chips, surface cracks or scuffs to facing bricks or blocks are within acceptable limits.

And remember to ensure that the bricks and blocks are stored in neat stacks (Figures 14.5 and 14.6).

Cement and Lime

Check that cement and lime are of the type specified and that they come from a consistent source, especially if used for facing brickwork. Cements from different sources may result in lighter or darker mortars, causing the brickwork to appear uneven in colour.

Figure 14.5 Stacked bricks.

Figure 14.6 Stacked blocks.

Sand

Check that sand is clean and, if for use in facework, from a consistent source. Check that it is certified by the supplier to be as specified. Different sands can result in distinctly different mortar colours.

Pre-mixed Lime/Sand and Ready-to-use Retarded Mortars

Check that the delivery note refers to the mortar as specified and visually check for colour consistency.

Damp Proof Course

Check that the damp proof course material type, width, weight or thickness and any standard and special pre-formed cavity units are as specified.

Wall Ties

Check delivery notes and where possible the ties (Figure 14.7) against the specification for:

- Type: vertical twist, double triangle, butterfly and whether suitable for use with partial-fill cavity insulation if specified.
- Material: the specified material, including the gauge or thickness of any galvanising.
- Length: the length of cavity wall ties must be such that a minimum of 50mm can be built into each leaf to provide an adequate structural connection.

Figure 14.7   Wall ties.

Personal Protective Equipment

Personal protective equipment must be supplied by your employer free of charge and you have responsibility as an employee to look after it and use it whenever it is required.

Information

For more information about materials, visit the websites of the better-known manufacturers and suppliers.

Quick Quiz

1. What is concrete composed of?

2. List three types of bricks.

3. Name the three categories of frost resistance for bricks.

4. What is lime mainly used for?

5. List five different types of cement.

CHAPTER 15

Cavity Walls

THIS CHAPTER RELATES IN THE LEVEL 1 DIPLOMA TO UNIT CC1016K AND UNIT CC1016S.

Cavity Wall Construction

Walls in exposed positions subjected to driving rain are likely to become damp internally by moisture percolating through either the joints, the bricks or the shrinkage cracks that often occur between the joint and the bricks. Because of this possibility, precautions are taken to prevent it; one of the most common measures usually adopted is the use of a cavity form of construction.

Cavity Walling

The cavity wall (Figure 15.1), which more directly concerns the brick-layer than any other, consists of two walls with a 50 mm minimum sized cavity or wider depending on the type of insulation specified between them, the outer wall being usually 102.5 mm thick and the inner wall sufficiently strong to carry loads imposed upon it. The usual size for this wall in ordinary house construction is also 102.5 mm.

Cavity walling has many advantages over solid wall construction; the following points should be noted:

- As there is no contact between inner and outer walls except at the wall ties, which are of impervious material, the possibility of moisture penetration is reduced to a minimum.
- If desired, use may be made of soft, porous facing bricks without causing the inside of the building to become damp.
- Common bricks or blocks may be used for the interior and, since no transverse bonding is required, a considerable saving will be made in the cost of facing bricks, owing to the absence of headers.

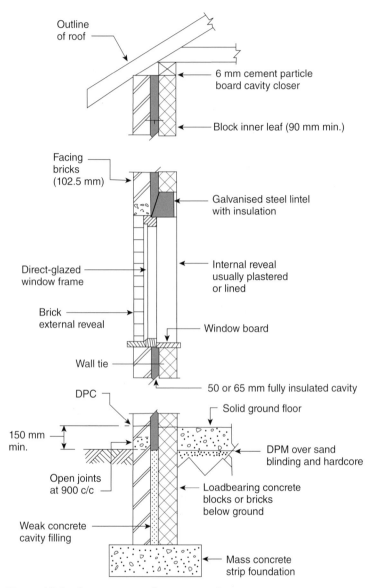

Outline of roof

6 mm cement particle board cavity closer

Block inner leaf (90 mm min.)

Facing bricks (102.5 mm)

Galvanised steel lintel with insulation

Direct-glazed window frame

Internal reveal usually plastered or lined

Brick external reveal

Window board

Wall tie

50 or 65 mm fully insulated cavity

DPC

Solid ground floor

150 mm min.

DPM over sand blinding and hardcore

Open joints at 900 c/c

Loadbearing concrete blocks or bricks below ground

Weak concrete cavity filling

Mass concrete strip foundation

Figure 15.1   Section view of a cavity wall.

● Greater consistency of temperature is maintained inside the building since no direct contact is made by the inner walling with the outside air, and the layer of air between the two parts of the wall is an excellent non-conductor.

The bond used for each skin or leaf of a cavity wall is usually stretcher bond; any other arrangement necessitates the use of bats, which may easily constitute a source of trouble unless cut accurately to size.

Damp Proof Courses

A damp proof course (DPC) is a layer of impervious material used to prevent the penetration of moisture, by absorption, to the inner parts of a building. Its introduction is necessary because building materials such as brick, stone and concrete are relatively porous.

There are several positions in a building where moisture may easily penetrate if provision is not made to prevent it, but it is generally accepted that the most vulnerable position is where the building makes contact with the ground, for at all times the ground contains a certain amount of moisture.

To prevent the moisture rising through the brickwork and reaching into the interior of the building, a horizontal layer of impervious material is built into the wall as the building proceeds. Extreme importance is attached to the placing of this layer, which is usually referred to as a damp proof course or DPC; its position is governed by the Building Regulations, which usually require it to be not less than 150 mm above ground level and beneath the floor timbers or solid floor.

For ordinary domestic work (i.e. house construction), a horizontal DPC is all that is necessary to keep the base of a wall dry, but where rooms are below or partly below ground level a vertical DPC will also be required.

Key Points

- Bed flexible DPC on fresh, smooth mortar.
- Lap DPC a minimum of 100 mm.
- Do not cover the exposed edge of DPC with mortar or render.
- Do not allow DPC to project into cavities.
- Extend DPC cavity trays beyond the ends of lintels, and ensure that stop ends are securely fixed.
- Ensure provision of sufficient weep holes.
- Use the correct type of DPC for the application

In addition to the DPC near ground level, the Building Regulations generally require a layer of concrete to be placed over the whole of the ground contained within the external walls. While this cannot be justly referred to as a DPC in the true sense, it assists in the prevention of dampness entering a building and reduces the risk of vegetable growth taking place or ground air penetrating into the building. This

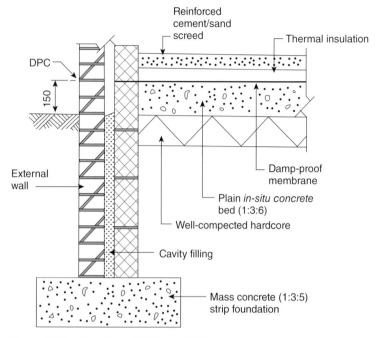

Figure 15.2 Position of DPC and DPM.

concrete, which is known as oversite, must be at least 150 mm thick if spade finished or 100 mm thick if properly laid on a bed of hardcore or similar material. (See Figure 15.2.)

Damp Proof Membrane

A damp proof membrane (DPM) or barrier is a continuous layer of damp-resisting material, the chief object of which is to protect the super-structure of a building against dampness.

The damp proofing of a wall near its base is undertaken to prevent the rise of dampness, which may be drawn from the ground, into the brick or blockwork forming the foundation walling of the building. The placing of a suitable DPM horizontally in a wall at a level of 150 mm above the adjoining ground level usually fulfils the above requirements. (See Figure 15.2)

Since it is essential to protect floors and floor timbers against dampness, it may be necessary to provide DPMs at different levels and in vertical positions around the building.

Material for use in DPMs should be permanently impervious to moisture and be durable. When placed in the wall or floor, they should be capable of resisting the loads imposed on them.

Lintels and Wall Ties

Use of Lintels

Lintels are generally introduced into the construction of a cavity wall to bridge over an opening at a required level and support the wall above, thereby transferring its weight to the piers on which the lintel rests. The use of a lintel is the simplest method of carrying a load over an opening, the lintel being made usually of steel (Figure 15.3).

Use of Wall Ties

In order to obtain the necessary stability, masonry walls are tied together with metallic or plastic wall ties (Figure 15.4) at suitable intervals throughout the length and height of the building.

The maximum spacing of the ties is governed by the Building Regulations, which state that they must not be placed more than 900 mm apart horizontally and 450 mm vertically, and must be staggered (Figure 15.5).

The purpose of a wall tie is to tie the two leaves of masonry together. Pushing the ties in displaces the mortar, which will not grip the tie effectively.

Note

It is a requirement of the Building Regulations that adequate means of support for the superstructure should be provided over every opening or recess in a wall. Build-in lintels with adequate bearings as specified or recommended by the manufacturer but never less than 100 mm. Bed them on mortar on full blocks, not short lengths of cut blocks. Whatever lintel is chosen, it is important that the correct type be used for the purpose intended.

Tip

Lintel manufacturers and suppliers are always willing to give advice on the use and application of their products.

Remember In a main wall, ties are required every six courses of brickwork or every two block courses.

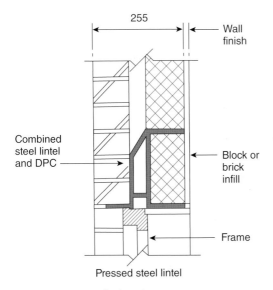

255

Wall finish

Combined steel lintel and DPC

Block or brick infill

Frame

Pressed steel lintel

Figure 15.3 Section view of a lintel.

(a) (b)

Figure 15.4   Wall ties.

Building-in Wall Ties

- Lay ties on a bed of mortar as bricklaying proceeds. Do not push the ties into the joints afterwards.
- Bed ties into the joint by at least 50 mm. They should be horizontal but preferably sloping slightly downwards to the outer leaf. Drips should be in the centre of the cavity and pointing downwards.
- If the bed is less than 50 mm, the wall will be weakened, especially against lateral or sideways load.
- Build-in ties staggered at alternate courses, unless otherwise specified, at the spacings shown in Figure 15.5.

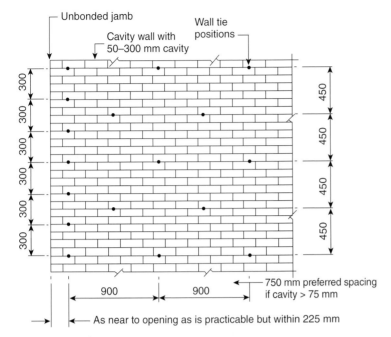

Figure 15.5 Wall tie spacings.

Heights of Cavity Leaves

When building with vertical twist ties, do not raise either leaf above the other more than the vertical spacing between horizontal rows of ties. This avoids the risk of broken bed joints and displaced units as ties are bent up or down to meet courses in the second leaf, which are out of level.

When building with less rigid ties, for example butterfly or double triangle type, do not raise either leaf more than 1.5 m above the other.

There are a wide variety of wall ties available for traditional brick and blockwork walls. They come in three types:

- heavy-duty wall tie for buildings of any height
- general purpose wall tie for buildings not greater than 15 m in height
- light-duty wall tie for buildings not greater than 10 m in height.

The ties described above are the most commonly available and the most frequently used.

Insulation

One of the aims of the Building Regulations is to improve the energy efficiency of all new buildings and existing buildings, when they can be extended or altered. This can easily be achieved with the range of insulants available (Figure 15.6).

Cavity insulation can be classified under three headings:

- full fill
- partial fill
- injection (after construction).

All these methods satisfy the requirements of the Building Regulations.

The recommended masonry cavity wall system is full fill, or built-in slabs. These systems not only provide the best U-value to wall width ratio but are also the lowest in cost.

Full-fill Insulation

This can be achieved by building in insulation bats as work proceeds or by filling the cavity with foam or granules on completion of the work.

Care is required when considering a fully filled cavity, since this can increase the likelihood of water penetration. When choosing an insulant for this application, reference should be made to the relevant British Standards.

Partial-fill Insulation

In partially filled cavity wall construction, a clear cavity of not less than 50 mm must be maintained in order to avoid bridging and to prevent the penetration of wind and rain.

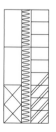
Section of cavity wall with full fill insulation

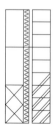
Section of cavity wall with partial fill insulation

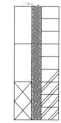
Section of cavity wall with post injected insulation

Figure 15.6 Types of insulation.

In order to accommodate the insulation and provide the required residual cavity, longer wall ties may be required together with special clips to fix the insulant securely to the inner leaf.

Insertion of an insulant within a cavity does not affect the durability of the external brickwork, but to reduce the risk of rain penetration to the internal skin, mortar joints should be completely filled, using only curved recessed or weather struck joints.

Injection (After Construction)

This is where the insulation is injected into the cavity after the main structure of the building is complete. Holes are drilled into the inner walls at about 1m centres and the insulation is pumped into the cavity.

Personal Protective Equipment

When working with bricks and blocks, make sure you always wear appropriate personal protective equipment (PPE), for example boots, safety helmet, gloves, goggles and face mask.

Appropriate eye protection equipment and dust suppression or extraction measures should be provided when mechanically cutting or chasing out brick- and blockwork.

Setting Out Cavity Walling

One of the main purposes of setting out brickwork for cavity walling is to create a matching and balanced appearance of bricks, particularly at the reveals on either side of door and window openings and ends of walls.

In this section only stretcher bond, also known as half-bond, is considered, although the basic principles will apply whatever bond is used.

Coordinating Size

Face brickwork should be set out before the bricklaying begins using as a unit dimension the coordinating size of the brick, that is one brick length (215mm) + one 10mm joint = 225mm.

Design

Wasteful cutting can be avoided and brickwork appearance enhanced if the overall lengths and heights of walls and door and window openings are all multiples of the brick unit size + one 10mm joint.

Perpends

Perpends (or perpendiculars) are the positions of vertical joints between the bricks. Their location should be decided at ground level. The verticality of perpends is visually important and the plumbing of every fourth or fifth perpend will produce satisfactory results.

Reveals

These are the sides of window and door openings, and the position of the reveal bricks should be identified when setting out the first few courses. This ensures unbroken perpends for the full height of the wall.

Reverse Bond

This is where the end bricks in a given course are showing a stretcher face at one end of the panel and a header face at the other. It can also apply at either side of an opening containing a half-brick size dimension in its width and where broken bond and brick cutting may be considered unacceptable. It is unlikely to be acceptable if reveal bricks of a contrasting colour are used as a decorative feature.

Building Cavity Walling

The purpose of cavity walls is to ensure that the inner leaf of the wall remains dry and that no moisture penetrates to the inside of the building.

The most efficient method of constructing cavity walling is to a line. The line is attached either to a raised corner or to corner profiles. It is important to bring the cavity wall up evenly to impose balanced loads onto the foundation.

The brick courses should be gauged at 75 mm per course but sometimes course sizes may vary slightly to accommodate window or door heights. Four courses of brickwork with a 10 mm bed joint equals 300 mm. This is known as standard gauge. This allows blockwork to run level with the brickwork on every third brick course.

For greater accuracy when building corners, they should only be raised between six or eight courses in height at any one time.

Damp Proof Barriers

The commencement of the cavity 150 mm below the DPC allows any condensed moisture in the cavity to drain below the level of the DPC before passing out through small weep holes (Figure 15.7) left in the form of open cross joints at intervals along the wall, and a space in which any mortar unavoidably dropped down the cavity during the

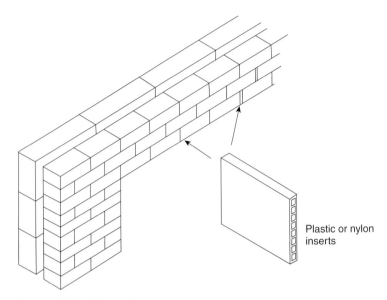

Plastic or nylon inserts

Figure 15.7 Weep holes.

process of building may collect. Great care must be taken to see that this mortar does not rise above the DPC or the effect of the cavity will be nullified.

Horizontal Damp Proof Course

The position of the DPC in relation to floor levels and the external ground level will be as for ordinary walling but, because of the necessity of maintaining the cavity below the DPC, it must be in two sections, one for the inner wall and one for the outer wall.

Damp Proof Course to Reveals

Since it is necessary, in cavity walling, to avoid all contact between the two walls, a vertical DPC must be introduced when sealing the cavity at window and door reveals. The manner in which this is done depends upon the general finish to the openings (Figure 15.8 shows two commonly used methods). Besides breaking contact between the two walls, it also seals off the back of the frame from any dampness existing in the cavity.

Damp Proof Course to Sills

As a protection against rotting of the material, a horizontal DPC is included under all timber sills (sometimes spelt 'cills'), a flexible type being normally used, such as polythene, examples of which are illustrated in Figure 15.8.

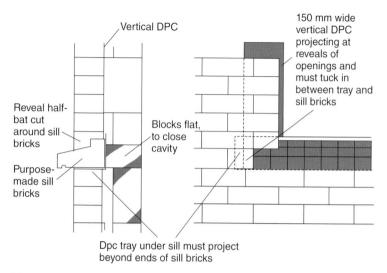

Figure 15.8 DPC to reveals and sills.

Damp Proof Course to Heads of Frames

The construction over window and door frames usually consists of a lintel to carry the brickwork over the opening, but no direct contact should be made between the two, a cavity tray being built across the cavity as illustrated in Figure 15.11. The cavity tray should extend lengthwise beyond the frame, 150 mm on either side, in order that any condensed moisture will drain beyond the sides of the frame.

Face Bonds

The bond used for each skin or leaf of a cavity wall is usually stretcher bond, any other arrangement necessitating the use of bats, which may easily constitute a source of trouble unless cut accurately to size. If cut short they are a weakness, while if too long they may possibly bridge part of the cavity.

Broken Bond

This is the introduction of cut bricks into a length of wall which, if properly set out, will maintain a satisfactory appearance and achieve a minimum quarter bond.

It is sometimes the case that the length of a given wall will not always be a multiple of brick dimensions or, alternatively, will not always permit the general sequence of the bond to be followed, particularly in the case of Flemish bond.

Half bat in wall

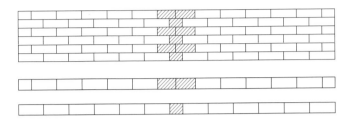

Three-quarter bat in wall

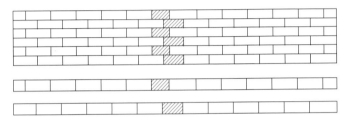

Figure 15.9 Broken bond.

From this it follows that a cut brick is sometimes necessary to complete a course or a change in the general arrangement is required, either of these alternatives being known as broken bond (Figure 15.9).

Keeping Cavities Clean

Up to ground level, the brickwork will be similar to that for ordinary walling but, beyond this, certain precautions are necessary, the most important being to keep the wall ties and cavity free from mortar droppings.

At all changes of direction and elsewhere where convenient, bricks should be left out at the base of the cavity so that any dropped mortar may be removed at the end of each day. As previously mentioned, this is essential to prevent the cavity becoming closed and moisture travelling across.

To reduce to a minimum the amount of mortar dropping to the bottom of the cavity, wooden battens are placed across the wall ties (Figure 15.10) and raised as each row of ties is required. Any mortar remaining on the ties may be removed with a short length of batten before the work proceeds. Mortar accumulating on ties often forms a bridge across the cavity and, for this reason, every endeavour must be made to keep them clear.

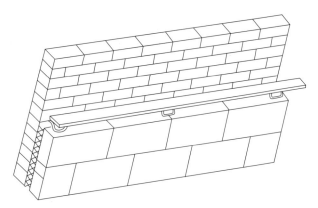

Figure 15.10   Cavity battens.

DPCs over window and door openings must be given particular attention by leaving out bricks to facilitate the removal of mortar droppings that easily accumulate at these points.

Whenever it is necessary temporarily to leave out a brick for cleaning purposes, it will be found convenient first to bed the bricks in sand; this procedure makes subsequent operations much easier and ensures that the correct bond is maintained.

Resistance to Moisture

Prevention of liquid water penetration from the outer to the inner leaf is one of the major considerations when designing cavity walls. The selection of appropriate materials and pointing methods for the outer leaf is crucial.

Cavity Trays

Cavity trays (Figure 15.11) should be provided:

- at all interruptions of the cavity, such as lintels and sleeved vents and ducts;
- above insulation that stops short of the top of the wall.

Cavity trays should rise at least 140 mm within the cavity, be self-supporting or fully supported with joints lapped and sealed. Stop ends should be provided to the ends of all cavity trays. Weep holes should be provided at not more than 900 mm centres to drain each cavity tray.

Condensation

In cavity wall construction, it is necessary to ensure that there is no danger of condensation forming within the inner leaf.

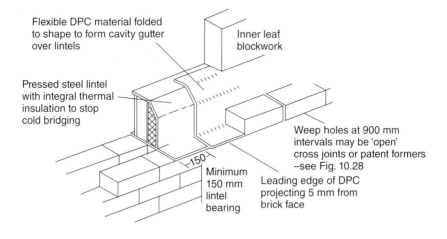

Flexible DPC material folded to shape to form cavity gutter over lintels

Inner leaf blockwork

Pressed steel lintel with integral thermal insulation to stop cold bridging

Weep holes at 900 mm intervals may be 'open' cross joints or patent formers –see Fig. 10.28

150

Minimum 150 mm lintel bearing

Leading edge of DPC projecting 5 mm from brick face

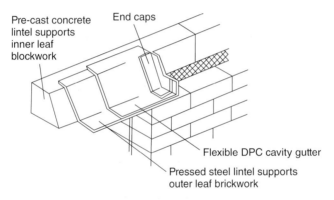

Pre-cast concrete lintel supports inner leaf blockwork

End caps

Flexible DPC cavity gutter

Pressed steel lintel supports outer leaf brickwork

Figure 15.11   Cavity trays.

Condensation may have a detrimental effect on the thermal perform-ance of a structure or cause damp on the inside. Unfaced mineral wool products, being 'vapour open', offer virtually no resistance to the passage of water vapour.

Resistance to Fire

Open cavities must be stopped to prevent the passage of fire. This is required at specific intervals and the cavity stop has to provide at least 30 minutes fire resistance. If the cavity is fully or partially filled and is built in accordance with Diagram 32 of approved Document B from the Building Regulations (a copy of which you should be able to get from your tutor or the local library), fire barriers are not required.

Solar Gain

Buildings oriented to face south can benefit from 'solar gain'. The thermal mass of parts of the structure heats up during the day and contributes to the night-time heating of the building. On top of the heating benefits, this minimises the need for artificial light during the daytime.

Working to a Specification

During your workshop training and on site, you will need to work to the required specification as described below.

Building Cavity Walling

Build the straight cavity wall shown on your workshop drawing in facing bricks to the front and blocks to the rear. Build two courses of common brickwork up to the DPC on the internal wall. Joints are to be half round on the front face and left flush from the trowel on the rear face.

Note

Check the information provided on your workshop drawing and industrial standards marking sheet before commencing.

Information

For more information about cavity walling, visit your local or college library and borrow textbooks on the subject.

Quick Quiz

1. List four advantages cavity walling has over a solid wall.
2. How far should a DPC be above ground level?
3. List three types of wall tie.
4. What is the minimum bearing for a lintel?
5. List the three classifications of insulation.

CHAPTER 16

Setting Out and Levelling

THIS CHAPTER RELATES TO UNIT CC1017K AND UNIT CC1017S.

Working Drawings

Working drawings are critical to construction work. It is essential that everyone in the building team is able to understand and interpret them as appropriate to their needs. To do this requires good spatial skills, an understanding of scale and the ability to interpret symbols and metric measurement.

Without drawings and plans, you and the building team would have no idea of what to build and how it should look.

Working drawings can be classified as:

- location drawings (include site and block plans)
- component range drawings
- assembly or detailed drawings.

> **Note**
>
> Once you have done an activity described in this chapter, ask your tutor to look over it to see whether you have reached the required standard for that specific area (what is known as the industrial standard).

Location Drawings

Site plans give the position of the proposed building and the general layout of the roads, services and drainage on site. Block plans identify the proposed site by giving an aerial view of the site in relation to the surrounding area (Figure 16.1).

Component Range Drawings

These drawings show the basic sizes and reference system of a standard range of components produced by a manufacturer. This is useful in choosing components suitable for a job (Figure 16.2).

Assembly or Detailed Drawings

These include all the information required to produce a given component. They show how things are assembled and what the finished item will look like on completion (Figure 16.3).

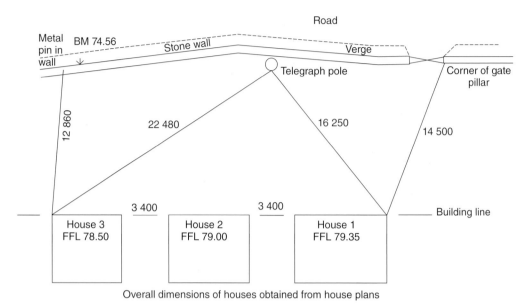

Figure 16.1 Location drawing.

Scales Commonly Used

- Block plans – Scales 1:2500 and 1:1250
- Site plans – Scales 1:500 and 1:200
- General location drawings – Scales 1:200, 1:100 and 1:50
- Range drawings – Scales 1:100, 1:50 and 1:20
- Detail drawings – Scales 1:10, 1:5 and 1:1
- Assembly drawings – Scales 1:20, 1:10 and 1:5

These scales mean that, for example, on a site plan drawn to 1:200 scale, 1 mm on the working drawing would represent 200 mm on the actual building. (See the section on scale drawings below.)

Keep Your Working Drawings Safe

In the harsh environment of a construction site or brick training workshop, working drawings can be easily damaged. This can lead to mistakes being made that can be expensive to rectify.

If working drawings are continually folded, the fold lines will become illegible and vital information may be lost. To avoid this, working drawings should be fixed to a board and covered with a transparent film, which will also protect them from dust and adverse weather.

Using special bricks

The drawing below (and the photograph opposite) indicates a few ways in which standard specials can be used to provide interesting brickwork features within a property. Of course, the design possibilities and permutations are infinite.

Successful detailing not only gives a unique aesthetic appearance but ensures the wall will perform satidfactorily for many years to come.

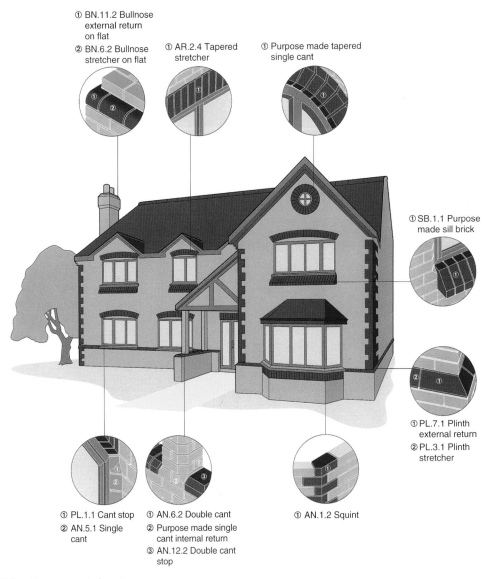

① BN.11.2 Bullnose external return on flat
② BN.6.2 Bullnose stretcher on flat

① AR.2.4 Tapered stretcher

① Purpose made tapered single cant

① SB.1.1 Purpose made sill brick

① PL.7.1 Plinth external return
② PL.3.1 Plinth stretcher

① PL.1.1 Cant stop
② AN.5.1 Single cant

① AN.6.2 Double cant
② Purpose made single cant internal return
③ AN.12.2 Double cant stop

① AN.1.2 Squint

Figure 16.2 Component drawing.

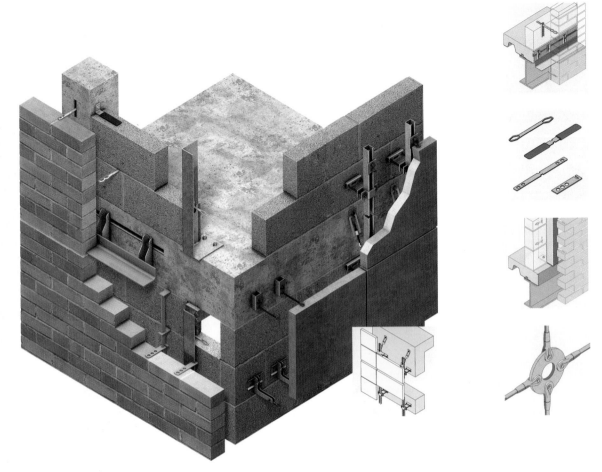

Figure 16.3 Assembly drawing.

Do not leave working drawings in the open for long periods as the process used to reproduce multiple copies causes those copies to fade quickly if exposed to sunlight.

Scale Drawings

Drawings used in construction are normally drawn to scale. This means that every measurement is in proportion to the real thing, for example:

- 1:1 = same size as the object
- 1:5 = 5 times smaller than the object
- 1:10 = 10 times smaller than the object

- 1 : 20 = 20 times smaller than the object
- 1 : 50 = 50 times smaller than the object
- 1 : 100 = 100 times smaller than the object

> *Remember* On working drawings, the second number in a scale tells you how many times bigger than the drawing the actual building will be.

Plans and Elevations

A working drawing shows what a building will look like and how it will be built. In order to give enough information, it is necessary to draw the building from several different view points. These different views are often on the same sheet (Figure 16.4).

- **Front elevation** A view of the front of the building
- **Side elevation** A view of the side/s of the building
- **Rear elevation** A view of the back of the building
- **Cross-section** A view of the building as though it has been sliced down from roof to floor
- **Plan view** Shows the building directly from above

> ### Tip
>
> The notes on a working drawing tell you what type of building is to be constructed and where it is to be built.

> ### Words and Meanings
>
> Elevation – A scale drawing showing the vertical image of a building.

Interpreting the Working Drawing

On the working drawing, you will find all the measurements you need, for example length and height of the walls, sizes of any door or window openings, height of the damp proof course.

Great care is taken to show all measurements. If this is not done, it is impossible for you to calculate the amount of materials you require and get the activity done at all. Guesswork is not a word used in bricklaying.

> ### Try this Out
>
> Using the plan view from a working drawing supplied by your tutor, draw up a list of the materials and their quantities required to build the workshop activity shown in the drawing.

Measurement

Accurate measuring is a critical skill in construction. Working drawings are drawn to show measurements to the exact millimetre. It is essential that you set out the measurements that you have taken from the working drawing precisely.

You will measure out your work with tapes and steel rules. These measuring tools often have imperial measurements along one edge, shown as feet and inches. Metric measurements are shown along the

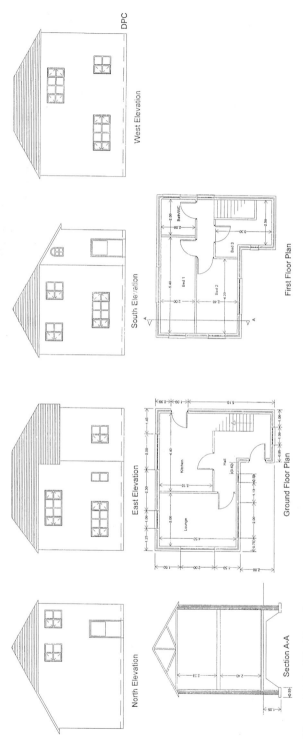

Figure 16.4 Working drawing.

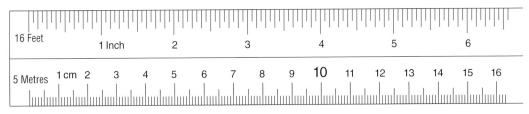

Figure 16.5 Tape measure.

other edge: metres (m) and centimetres (cm). The most important measurements used in bricklaying, millimetres (mm), are shown as the little lines between each marked centimetre (Figure 16.5). See Chapter 9 for more about metric and imperial measurements and for tips on how to convert between the two.

There are 10 millimetres in every centimetre. There are 100 centimetres in a metre. So to find out how many millimetres there are in a metre you have to multiply by 10, for example:

$$100 \text{ cm} \times 10 = 1000 \text{ mm.}$$

If you have 20 cm and you want to make this into mm, then you have to think of 20 lots of 10. It is:

$$20 \text{ cm} \times 10 = 200 \text{ mm.}$$

Tip

There are 1000 millimetres in a metre. To convert metres to millimetres multiply by 1000. For example 2 m × 1000 = 2000 mm

Remember If you want to measure in millimetres, you have to think and count in tens and hundreds.

Dimensional Accuracy

The accuracy of horizontal and vertical measurements is critical if work is to be completed to industrial standards. It is for this reason that during your training at college dimensional accuracy will play a major part in all assessments.

The construction industry does not use centimetres. All dimensions will be given in metres or millimetres and will be written in one of the following ways:

- 3148 or 3.148
- 900 or .900
- 2703 or 2.703
- 287 or .287 and so on

Before you begin setting out, make sure you understand clearly the divisions on the tape measure you are going to use. If you are unsure, ask your tutor for clarification.

Tape measures are provided with either a hook or a hook and ring to allow you to fix the tape over the edge of an object or to secure tape over a nail in the top of a peg to give greater accuracy.

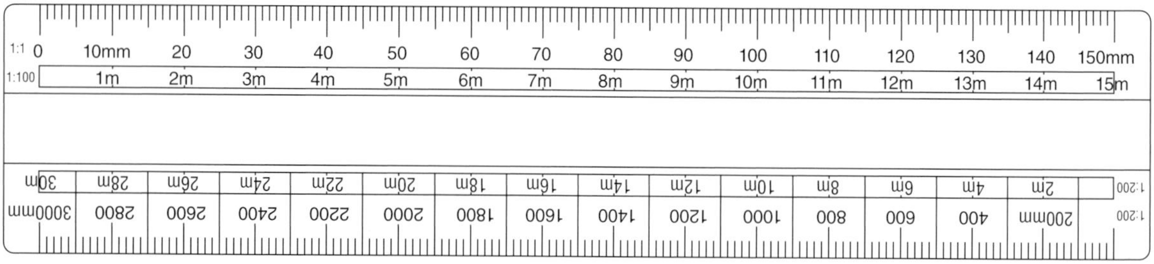

Figure 16.6 Scale rule.

The Scale Rule

Scale rules (Figure 16.6) are used to create drawings and are very useful for quickly interpreting dimensions. It can be very helpful to use a scale rule to deal with preparing and reading scale drawings. There are different scale rules that can be used depending on the scales you need.

For example, if you have measured something on a working drawing that reads as 50 mm and the scale used is 1 : 10, using the 1 : 10 scale section on a scale rule will show you that this is 500 mm in full size (50 × 10 = 500). Some other examples are:

- On a scale of 1 : 5, 10 mm represents 50 mm (10 × 5 = 50).
- On a scale of 1 : 10, 10 mm represents 100 mm (10 × 10 = 100).
- On a scale of 1 : 20, 30 mm represents 600 mm (30 × 20 = 600).

If a drawing uses a 1 : 1 scale, it is drawn on paper the same size as it is in real life.

Setting Out

Setting out requires the bricklayer to transfer measurements from working drawings to the building plot. There are implications for the whole structure if the measurements are incorrectly interpreted at this stage in the building. Knowing how to read and handle metric measurements is therefore an essential skill for the bricklayer.

Reporting Inaccuracies

It is very important that you report any inaccuracies that you discover to your supervisor or tutor as soon as possible, either verbally or in writing. Inaccuracies or mistakes made at any stage during the construction of a building can be very costly to rectify. A wrong measurement or wrong angle at this stage can result in large sections of work having to be taken down at great expense to the company.

Tip

Always look at the scale written on working drawings. You can then use the correct scale rule to read off unmarked measurements accurately.

Remember If you spot an inaccuracy, report it as soon as possible.

Checks on Equipment

Always be aware of the limitations of equipment and instruments. This applies to the control of dimensions as much as it does to the initial setting out.

Always test the equipment you are about to use to ensure their accuracy. Instruments are used on site, then go back to the stores and in many cases are rarely checked. The instruments are then re-allocated and may be in any condition. So always make sure the instrument is properly set up, or make allowances for any faults or inaccuracies found. If a fault is discovered, return the instrument to stores for adjustment or repair.

Site Clearance

The reason site clearance is done is to prevent plants, shrubs or trees from attempting to grow; this would exert pressure on the concrete and crack it. Also, once covered over, the vegetation contained in the soil will decay, causing voids to form below the concrete. Most importantly, however, site clearance allows TBMs to be placed on reasonably level ground so as to be seen by site personnel from all areas of the site. The depth of vegetable topsoil varies, and on some sites it may be necessary to remove 300 mm or more vegetable topsoil.

Locating Existing Services

Before undertaking any digging, for example trial holes, it is necessary to clearly identify the nature of services on the site and their actual position. The service providers will be keen to establish exact positions in an attempt to prevent damage to their pipes and cables. The organisations responsible for particular services should be invited to the site to help to establish the exact position, size and capacity of their supply and to resolve any uncertainty.

Datum Levels

TBM is the abbreviation for a temporary benchmark (Figure 16.7). A TBM is a known height point from which all levels are taken.

The TBM can be related to an Ordnance Survey benchmark. Ordnance Survey benchmarks appear on certain public buildings. The TBM can be surrounded by concrete for protection. A timber enclosure can be an additional protection.

Identification is often aided by the TBM being coloured. Levels are taken from the top of the TBM peg. Many different level points are related to the TBM.

Figure 16.7 Example of a temporary benchmark.

A datum height is a level point established in relation to a TBM. Datums are level reference points for various work activities, for example:

- damp proof courses
- invert levels
- oversite concrete
- foundations.

It is important that all setting out work is protected, as the setting out of a building is a time-consuming and exacting task and any infringement on the work could lead to errors and ultimately the building being set out incorrectly.

Transferring Datum Heights

Site levels are taken from the top of the TBM peg. Many different levels on site are related back to the TBM. Therefore, a datum height is a level point established in relation to a TBM.

Datum levels are transferred to the corners of the proposed building and other points by means of optical levels and a straight edge and spirit level.

Methods

Transferring datum heights may be defined as a method of expressing the relative heights of any number of points above or below some plane of reference called the datum. The site datum is transferred from the nearest ordnance benchmark on to the site by means of transferring levels. Some examples of transferring levels are described below.

Water Level

A water level is a length of hose with a transparent tube set in each end (Figure 16.8). The hose is filled with water, a cap or cork being provided to prevent spilling. Care must be taken to see that no air is trapped when filling.

Figure 16.8   Water level.

The level works on the principle that water finds its own level. It can be used over distances of 30m or so and is useful for marking a number of points quickly, especially around corners and obstructions.

Spirit Level and Straight Edge

This is a very basic method of transferring a level. The straight edge and level are rotated through 180 degrees at each levelling point to eliminate any error in the level or straight edge.

Method

- **Stage one:** Start from the TBM. Align a straight edge towards the intended datum point. Drive in a temporary peg at the full extent of the straight edge, level with the TBM. Check for level.
- **Stage two:** Repeat the process to extend the temporary levelling pegs until the final datum peg positioned is reached. For greater accuracy, reverse the spirit level and the straight edge at each intermediate levelling stage.
- **Stage three:** Datum pegs are located close to the corner profiles. Datum peg heights may be adjusted in relation to the TBM to suit work activity.

Optical Level

Optical levelling is another method of transferring levels, and is suitable for greater distances.

Setting up An Optical Level

The equipment and materials required are:

- optical level and tripod (Figure 16.9)
- levelling staff
- notepad and pencil

Method

Attach the level to the tripod. Using the footscrews, centre the fish-eye bubble. Sight on to the staff. Take a number of staff readings.

Cowley Optical Level

A Cowley, or Quickset, level is an optical levelling device that comprises the levelling instrument, a tripod and a target (staff). The instrument swivels through 360 degrees on a pin at the top of the tripod and is accurate up to about 30m.

Using the Cowley Level

The equipment and materials required are:

Figure 16.9 Optical level.

- Cowley level and staff
- bricks and mortar
- brick trowel
- notepad and pencil.

Follow the method described below:

- Set the Cowley level up in the prescribed manner. Bed two bricks at point A. take a level reading of the bricks and note this down.
- Bed two bricks at point B and C, as illustrated in Figure 16.10. Adjust the height of the bricks until they are level with point A.

LASER Level

The letters of LASER tell you what the equipment really does. They stand for Light Amplification by Stimulated Emission of Radiation. Most lasers used in the construction industry are of the helium/neon gas type that produces a polarised beam in the form of an intense red light.

Setting up A LASER Level

The equipment and materials required are:

Figure 16.10 Cowley level.

Figure 16.11 LASER level.

- LASER level and tripod
- levelling staff
- notepad and pencil.

Method

The level is set up using the fish-eye bubble as for the optical level. The level is then turned on and a LASER beam is sent out on a level plane. A special target receiver is then used to transfer this level to a target at the position where the level is to be taken and set out.

The Building Line

The building line is a line fixed by the Local Authority Planning Department as a limit for building towards a road.

Before any setting out can be done, the position of the building on the site must be determined, this being found from the block plan which refers only to the position of the building and gives no detailed information as to the building itself.

The position of the building line is usually indicated by a dimension from a fixed line or point such as an existing building or road. This line

Figure 16.12 Pegs and builder's square used for setting out.

is marked on the site by pegs of 50 mm × 50 mm timber driven into the ground at the extreme corners on the front of the building, nails or saw cuts being used to indicate the exact position on top of the pegs.

Dimensions for the proposed building are taken from the drawings. Check the drawings so that overall dimensions compare with individual dimensions. Work to marked dimensions only. Do not scale from drawings. Sometimes, only the overall dimensions are shown. Opening widths may not have a dimension shown.

The front corner position of the building is often marked on the kerb. The overall width of a plot or the boundary can also be marked on the kerb (Figure 16.13).

Check the information marks on the kerb. Different methods of marking may be used.

A building can be related to the building line (Figure 16.14) in different ways. For example, a building can be:

Figure 16.13 Kerb markings and builders square.

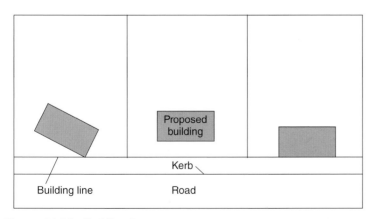

Figure 16.14 Building line.

- angled to the building line
- behind the building line
- on the building line.

In special cases, part of the building may be permitted to project beyond the building line, for example a porch.

Profiles and Ranging Lines

Profile boards indicate the positions of walls and foundations. Ranging lines are located onto the profiles, which trace out the alignment of walls and foundation trenches.

Overall wall and foundation trench widths are shown on the profiles. Pegs are positioned on the inside of the profile board cross pieces. This prevents the line tension pulling the cross piece from the pegs.

Profile boards are constructed from timber pegs and cross pieces. Approximate sizes of timber are:

- pegs 50 mm × 50 mm
- cross pieces 100 mm × 20 mm section.

Profiles must be rigid. Pegs must be firm in the ground. Nails should be staggered diagonally on the cross piece for maximum rigidity and left protruding.

The length of cross piece must be sufficient to contain the information required, namely wall and foundation widths (Figure 16.15).

Two profiles are required at each corner. Profiles are also necessary for any load-bearing internal partition walls.

Note

For accuracy, it is important that all profiles be at the same level.

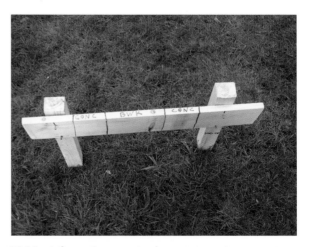

Figure 16.15 Information required on a cross piece.

Working Space

In order to obtain the setting out points on profiles after excavation commences, profiles are erected at points where they will not be subsequently disturbed, the usual position being about 1 m away beyond the projection of the foundation; the space allows working room for operatives. The cross pieces should be level.

Establishing Ranging Lines

Stage 1

Position a cross piece clear of the corner peg and at least 1 m clear of the excavation. Make allowances on cross pieces for the position of the outer edge of the foundation trench excavation. Mark on profile board. The ranging line should be just touching the nail (Figure 16.16).

Stage 2

Drive in two pegs to accommodate the cross piece.

Stage 3

Place cross piece into position to determine position of nails.

Figure 16.16   Cross piece and ranging line.

Figure 16.17 Profile boards in place.

Stage 4

Partly drive nails into each end of the cross piece. Position the nails diagonally to provide rigidity.

Stage 5

Use a sledge hammer to provide a support to the pegs when driving in the nails. Drive in one nail.

Stage 6

Level the cross piece using a spirit level. Drive in the remaining nails. Do not drive the nails fully home. The nail heads left protruding help when dismantling the profiles.

Repeat the process at all corners. The result should look like Figure 16.17 with eight profile boards in place.

Try this Out

Describe methods of accurately locating walling and trench positions onto single wall and corner type profiles using information from block plans, site plans and drawings.

Setting Out Corners

The Builder's Square

The builder's square is used for the setting out of right angles on site. A builder's square is usually made from wood. It is braced to maintain its squareness and one side is longer than the other. When first made, a square's accuracy can be guaranteed, but on site and after exposure to the weather and possibly misuse its accuracy should not be relied on.

Most builder' s squares are made of 75mm × 30mm timber half jointed at the 90 degree angle with a diagonal brace, tenoned or dovetailed into the side length (Figure 16.13).

When setting out with a builder's square, accuracy depends on lining up the ranging line with the side of the square. Greater accuracy can be achieved when the sides of the square are increased in length.

Optical Site Square

This is an optical instrument used for setting out right angles on site. The site square contains two telescopes set permanently at 90 degrees to each other (they can be adjusted vertically to enable fixing points

Figure 16.18 Wooden pegs.

to be located at convenient points) and a tripod with adjustable legs and a steel rod, which allows the instrument to be set up over a fixed point, for example a saw mark on a profile or a nail in a peg (Figure 16.19).

Figure 16.19 Site square.

Once the position of the corner has been established, it is marked by driving in a peg. For greater accuracy, a nail is driven into the top surface of the peg establishing the exact position of the corner.

The site square is then positioned above this using the plumb or datum rod for accuracy.

The site square will give a range from 2 to 90 m.

Method

Set out the base line of the building. Set up the site square over the corner peg with a nail slightly protruding from the top of it. Line the site square up with the base line. Lock site square into position. Reposition second ranging line to line up with right-angled telescope. At this point, both lines will be at right angles to the base line.

The 3:4:5 Method

A right angle can be produced using the 3:4:5 method. A triangle constructed with sides measuring 3, 4 and 5 units will produce a right angle at A (Figure 16.20).

Progress Check

The task of setting out and checking angles using the 3:4:5 method should be carried out under the supervision of your tutor.

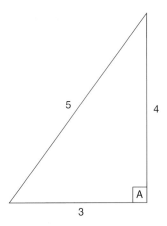

Figure 16.20 Creating a right angle using the 3:4:5 method.

Setting Out Walling

After the foundation of the structure has been completed and it is ready for bricklaying, it is necessary to transfer the wall profile to the foundation in order to locate and establish the corner profiles for wall construction.

The ranging lines fastened to the profiles represent walls and their locations.

Once the lines have been fixed, you can transfer the wall line to the foundation. To do this, you must use a spirit level or plumb bob to plumb down from the outside wall line to mark its position on the foundation.

When you feel you understand this procedure, ask your tutor to observe you whilst you carry out this operation.

Try this Out

Setting Out a Small Building

Method

Obtain a working drawing from your tutor. Refer to this drawing at all times, ensuring that you use the dimensions on it and no others. Do not use scaled measurements.

Determine the Building Line

Measure from any fixed point indicated on your drawing. The building line will be defined by the use of pegs with nails in the top and ranging lines strung between them. Figure 16.21 shows the wooden pegs placed on the boundary of the site. This is an excellent place to afford protection to the pegs and to assist later when transferring the ranging lines to the profile boards.

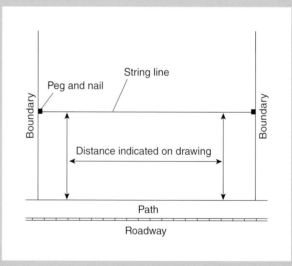

Figure 16.21   Building line and boundaries of site.

Set Out External Corners

In order to set out the external front corners you will need to determine the measurement from one of the site boundaries. The boundaries will be shown on your working drawing, usually on the block plan.

Method

Peg A is placed in position by measuring from the boundary of the site.

Peg B is placed in position by taking the width of the proposed building from the working drawing, as shown in Figure 16.22.

Setting Out Right Angles

In the first part of this chapter you practised setting out right angles using the builder's square and optical levels. In this part of the chapter you can choose either method for setting out. Figure 16.23 shows the working drawing with all four corners set out. Pegs A, B, C and D now establish the corners of the building.

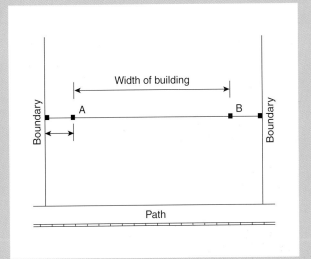

Figure 16.22 Width of proposed building and boundary.

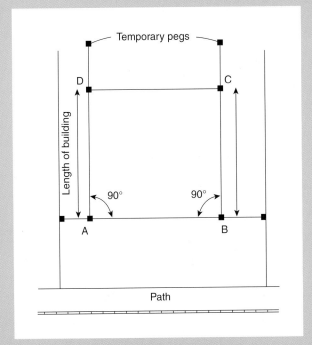

Figure 16.23 All four corners set out.

An alternative method of setting out right angles is to set out a right angle at point A and then complete the setting out by means of establishing parallel lines E–E and F–F, as shown in Figure 16.24.

Checking Diagonals

Before carrying out the operation of setting out the corners, check that your dimensions for width and length of the building are correct. Once you have done this you can check for squareness. To do this, measure the diagonals from A to C and B to D, as shown in Figure 16.25. For the building to be square, they should both be equal in length.

Before going on to the next part of this setting out exercise, make a final check of the overall dimensions.

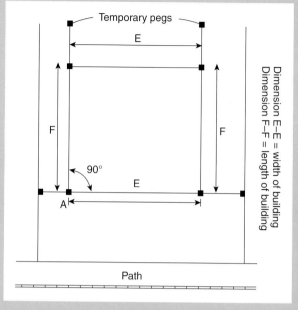

Figure 16.24 Setting out using parallel lines.

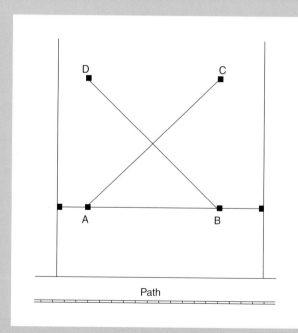

Figure 16.25 Checking of diagonals.

Transferring Ranging Lines on to Profiles

The profile boards and pegs can now be erected in order to carry the ranging lines clear of the excavation. The profiles should be extended approximately 1 m away from the excavation if you intend to hand dig them and some 5 m away if using a machine. Ensure you keep these distances the same for all four sides, as shown in Figure 16.26. This will assist you when you come to check the final dimensions.

The required setting out information, marked on each profile board, is illustrated in Figure 16.27.

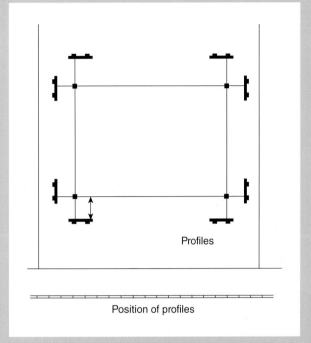

Profiles

Position of profiles

Figure 16.26 Site showing layout of profile boards.

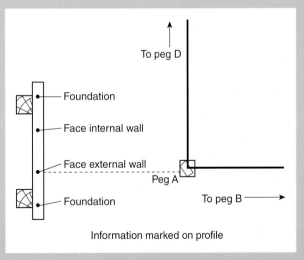

To peg D

Foundation

Face internal wall

Face external wall

Peg A

To peg B

Foundation

Information marked on profile

Figure 16.27 Profile board showing information markings.

Transferring External Dimensions

In order to transfer the external dimensions of the building from the pegs to the profiles, you must now extend the ranging lines through to the profile board and fix it on to a projecting nail, as shown in Figure 16.28. The remainder of the marks on the profile board can be measured from the external wall of the building.

After setting out the external walls, profiles can now be erected for the internal walls by using a series of parallel lines from the original square. Excavation trenches can now be marked out by spray paint or a line of lime, sand or cement.

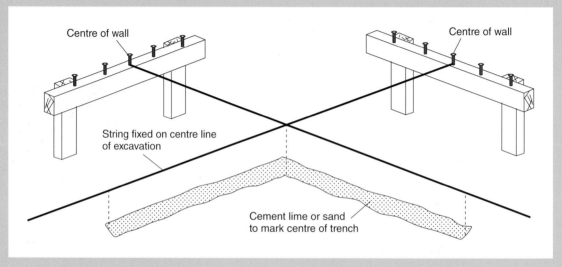

Centre of wall

Centre of wall

String fixed on centre line of excavation

Cement lime or sand to mark centre of trench

Figure 16.28   Attaching lines to profiles.

Industrial Standards

All practical work must be carried out to a specified quality. This quality is known as the industrial standard.

Reporting Errors

It is vitally important to follow correct procedures when reporting errors or other circumstances which may affect the programme of work. The programme of work is a device for setting and recording work carried out on site and anything that may slow or halt ongoing work needs to be known by your supervisor. Therefore, if you do come across errors, you must immediately inform your supervisor either verbally or in writing about them so as to enable him or her to deal with the matter in a satisfactory manner.

Remember Errors that affect the programme of work cost your company time and money.

Information

For more information about the setting out of buildings, visit the websites of the Construction Awards Alliance (CAA): www.caalliance.co.uk.

Building Control

For more information about building control, contact your local town hall, which will be happy to help you with your enquiries.

Surveying Equipment

For more information about surveying equipment, visit the websites of the better-known manufacturers and suppliers.

Quick Quiz

1. What do the initials LASER stand for?

2. What name is given to a temporary height point from which all levels are taken?

3. What do profile boards indicate?

4. Name three classifications used for working drawings.

5. List five scales used in construction.

INDEX